Digital Design Using Field Programmable Gate Arrays

Prentice Hall Series in Innovative Technology

Dennis R. Allison, David J. Farber, and Bruce D. Shriver *Series Advisors*

Bhasker	*A VHDL Primer*
Blachman	*Mathematica: A Practical Approach*
Chan and Mourad	*Digital Design Using Field Programmable Gate Arrays*
El-Rewini, Lewis, and Ali	*Task Scheduling in Parallel and Distrubuted Systems*
Johnson	*Superscalar Microprocessor Design*
Kane and Heinrich	*MIPS RISC Architecture, Second Edition*
Kehoe	*Zen and the Art of the Internet: A Beginner's Guide, Third Edition*
Lawson	*Parallel Processing in Industrial Real-Time Applications*
Nelson, ed.	*Systems Programming with Modula-e*
Nutt	*Open Systems*
Rose	*The Little Black Book: A Practical Perspective on OSI Directory Services*
Rose	*The Open Book: A Practical Perspective on OSI*
Rose	*The Simple Book: An Introduction to Management of TCP/IP-Based Internets*
Shapiro	*A C++ Toolkit*
Slater	*Microprocessor-Based Design*
SPARC International, Inc.	*The SPARC Architecture Manual, Version 8*
SPARC International, Inc.	*The SPARC Architecture Manual, Version 9*
Strom, et al.	*Hermes: A Language for Distributed Computing*
Treseler	*Designing State Machine Controllers Using Programmable Logic*
Wirfs-Brock, Wilkerson, and Weiner	*Designing Object-Oriented Software*

Digital Design Using Field Programmable Gate Arrays

Pak K. Chan
University of California, Santa Cruz

Samiha Mourad
Santa Clara University

PEARSON

ISBN 978-81-317-2440-8

10 9 8 7 6 5 4 3

Published by Pearson India Education Services Pvt.Ltd,CIN:U72200TN2005PTC057128.
Formerly known as TutorVista Global Pvt Ltd, licensees of Pearson Education in South Asia

Head Office: 7th Floor, knowledge Boulevard, A-8(A) Sector-62, Noida (U.P) 201309, India

Registered Office: *Module G4, Ground Floor, Elnet Software City, TS-140, Block 2 and 9, Rajiv Gandhi Salai, Taramani, Chennai 600 113, Tamil Nadu, India. Fax: 080-30461003, Phone: 080-30461060,* www.pearson.co.in, *Email:* companysecretary.india@pearson.com

Printed in India by Saurabh Printers Pvt Ltd.

Contents

List of Tables

List of Figures

Digital Design Using Field Programmable Gate Arrays

Chapter 1

Introduction

1.1 Overview

Digital circuit design has gone through several evolutions in the last few decades. The dramatic change in technology has radically changed the design process. Circuit components have evolved from individual transistors to very-large-scale integrated circuits (VLSI). Computer-aided design (CAD) tools have accelerated the design cycle. It is no longer necessary to assemble different components, to draw individual gates, or to draw polygons. Hardware description languages (HDL) are more acceptable now for system design on a hierarchical level. Automated logic synthesis tools are available to create circuits that are readily mapped into silicon. In addition to the rapid change in the technologies and techniques of VLSI design, the life cycle of modern products is becoming shorter than its design cycle. Thus the need for rapid prototyping is ever growing.

Programmable logic devices have facilitated the prototyping task by reducing the physical design phase to a few minutes. However, to optimize the use of these devices, it is necessary for designers to understand the technology and the architecture of the hardware in which the design is implemented. It is the intent of this book to familiarize designers with the different aspects of designs targeted to field-programmable gate arrays (FPGAs).

The first chapter explains why FPGAs are important in digital circuit implementation, and gives a preview of all the chapters in the book. In Chapter 2, the components of an FPGA are described using two main types of FPGAs. The design flow and the CAD tools used in the implementation with FPGAs are the subject of Chapter 3. Chapters 4, 5 and 6 describe the logic design principles and practices as applied to FPGAs with an emphasis on Xilinx devices. Physical design is covered in Chapter 7, where special approaches to the placement and routing of FPGAs are noted. Chapter 8 is on testing. The concepts explained throughout the book are illustrated by case studies in Chapter 9.

In this chapter, we first situate implementation with FPGAs among different design methodologies. We then give the benefits, limitations, and applications of FPGAs. In Section 1.5 we outline the steps followed in the design cycle. Design verification and testing and CAD tools are the topics of the last section.

1.2 Design Methodologies

The implementation of a digital system is not independent of the design style. Digital integrated circuits (ICs), may be realized in different technologies depending on their size and their role in the system. This depends also on the economics of the project.

Integrated circuits consist of transistors that are placed on the chip and are connected in a such a way as to realize the design. The locations and connectivity of the transistors are defined by several *masks*. A mask corresponds to one of the silicon compound layers that form the transistors and the interconnect layers.

Circuit implementations may be grouped into two main categories, fully custom and semicustom designs, as illustrated in the hierarchy shown in Fig. 1.1. The latter category itself consists of several approaches. These approaches have facilitated the design and manufacturing of application-specific-integrated circuits (ASICs).

ASICs can be defined as ICs designed for a particular application in low volumes or end use such as in telecommunication, automotive, and so on. They can also be contrasted to standard ICs such as microprocessors and memories that are used in a wide range of applications and are available "off the shelf."

1. **Custom ICs.** Custom ICs are created using unique masks for all layers during the manufacturing process. The user controls chip density with high utilization. Since the designer controls all stages of the chip layout, maximum design flexibility and high performance are possible. Consequently, only highly skilled and competent designers are engaged with such design methodology. Also, development time is long, and development costs are extremely high. For applications that require high volume, custom ICs provide a low-cost alternative. The high cost of design and testing can be successfully amortized over the high volume.

2. **Mask-Programmable Gate Arrays (MPGA).** The gate array implementation approach uses generic masks for all but the metalization layers, which are customized to the user's specifications [7]. The generic masks create an array of modular functional blocks as shown in Fig. 1.2. Modules of transistors are arranged in rows that are separated by fixed-width channels. User logic is implemented by patterning these transistors into logic functions and connecting the different modules. The design is usually facilitated by a cell library, making the designer's expertise less

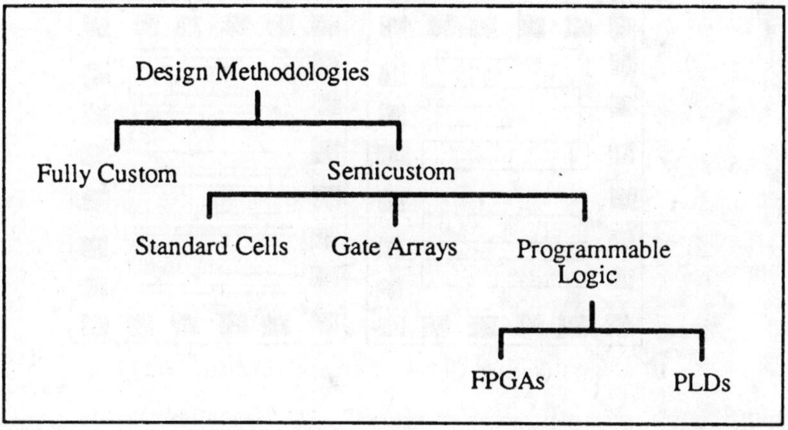

Figure 1.1: Design Methodologies.

critical than in the case of the full custom methodology. For the same reasons, MPGAs offer shorter development time and lower development costs than do custom ICs. A special class of gate arrays are channelless. They are known as sea-of-gates.

3. **Standard Cells.** In this approach, as in the case of MPGAs, the design task is facilitated by the use of predesigned modules (masks for these modules). The modules, standard cells, which are of the same pitch size, are usually saved in a database. Designers select cells from the database to realize their design. The cells are then placed in rows and interconnected. The routing is done within channels that may be of variable width. Placement and routing are done automatically almost removing the designers from the physical design process. Compared with custom ICs, circuits implemented in standard cells are less efficient in size and performance; however, their development cost is lower. Mixed standard cells and macros have also proliferated from standard cell design.

1.3 Field-Programmable Devices

Like gate arrays, field-programmable devices are prefabricated. However, the logic is implemented by electrically programming the interconnects and personalizing the basic cells, typically in the user's laboratory instead of a factory.

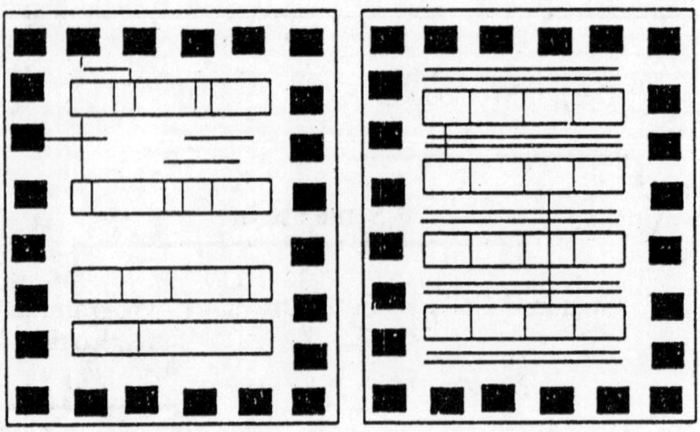

(a) Standard Cell (b) Gate Array

Figure 1.2: MPGA and Standard Cell Images.

The field-programmable devices have a variety of architectures. Implementing the design in programmable logic devices has the advantage of fast turnaround but limits the design flexibility, as we shall explain in Section 1.4. Development time and costs are significantly lower than are those of any other IC implementation. According to their architectures, we distinguish two main categories of user programmable logic devices: programmable logic devices (PLDs) and field-programmable gate arrays.

1. *PLDs* consist of programmable AND arrays (product terms) and fixed fanin programmable OR gates that are followed by flip-flops, as shown in Fig. 1.3. The outputs of the flip-flops can be fed back as input lines in the product terms. The product line can be connected to any combination of inputs. The connections are indicated by an "x" and are programmed by users to implement their designs. The connecting device may be a fuse as in the case of bipolar chips or a transistor. The transistor can be chosen to act as an open connection or to function normally as a switch [10]. PLDs are at the low-density end of field programmable logic devices. Their densities range from 1,000 up to 10,000 gates. Utilization varies with applications, but it is typically very low because of the rigid AND/OR architecture. Initially, PLDs used to be fabricated with bipolar technology; however, complementary metal-oxide semiconductors (CMOSs) are now more popular.

2. *FPGAs* combine the architecture of gate arrays with the programmability of PLDs. Some of the FPGA real estate is occupied by vendor logic to implement the field programmability feature of the FPGA, and a large

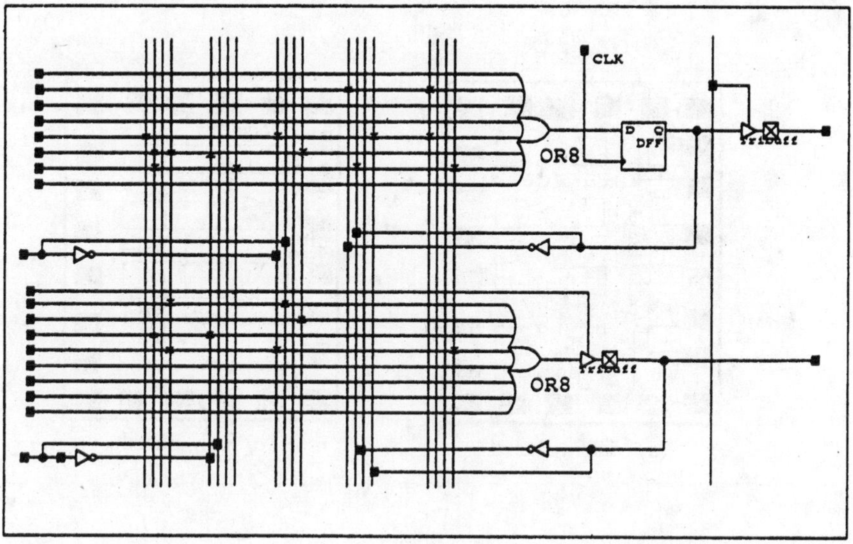

Figure 1.3: General Description of a PLD.

portion of the die area is for programmable routing. The number of gates typically available to the user varies from 3,000 to 10,000. An FPGA normally consists of several uncommitted logic blocks in which the design is to be encoded. The logic block consists of some universal gates, that is, gates that can be programmed to represent any function: multiplexers (MUXs), random-access memories (RAMS), NAND gates, transistors, etc. The connectivity between blocks is programmed via different types of devices, SRAM (static random-access memory), EEPROM (electrically erasable programmable read-only memory), or antifuse. The architecture of the chip depends on the fashion in which the blocks are arranged.

- They can form islands in a matrix with horizontal and vertical channels; Xilinx and QuickLogic devices are examples of this *matrix-based* architecture [15].
- They can form rows separated by routing channels like in a mask-programmable gate array; Actel devices are examples of this *row-based* architecture [2];

Both organizations are shown in Fig. 1.4. Details of the architecture and programming method of these devices will be covered in Chapter 2.

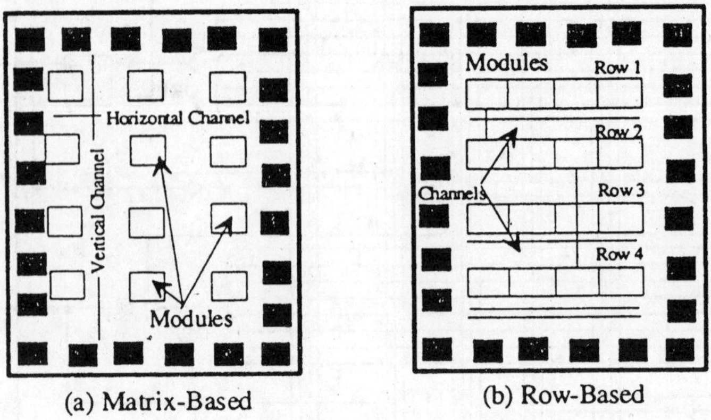

Figure 1.4: Main Architectures of FPGAs.

1.4 Why Field-Programmable Gate Arrays?

Recently, field-programmable gate arrays have enjoyed widespread use due to several advantages related to relatively high gate density, short design cycle, and low cost. They can be used in all applications that currently use small-scale integration (SSI), medium-scale integration (MSI), and PLDs (programmable logic devices). They also replace mask-programmable gate arrays in many applications that are limited to 10,000 gates and that do not require very high operational speed.

1.4.1 The Advantages and Limitations of FPGAs

Among the advantages of FPGAs are:

- Replacement of SSI and MSI chips
- Availability of parts off the shelf
- Rapid turnaround
- Low risk
- Reprogrammability (for some FPGAs)

Compared with SSI and MSI chips, FPGAs offer larger gate counts and more design flexibility. One FPGA can replace several SSI or MSI chips. If we consider a circuit constructed solely of NAND gates and its gate count is 1,000, we need about 250 each transistor-transistor logic (TTL) 7400 SSI chips to build it. The same circuit can, however, be replaced by 1 Xilinx chip [15], for example, an XC3020 or even an XC2064. FPGAs are more flexible because the logic does not have to be mapped in terms of standard SSI chips. Wire wrapping and soldering are also not required for a single-chip design, thus making it easier to realize appropriate engineering changes. While MSI chips have specific functions to which the design has to be mapped, FPGAs allow any random logic suitable for the designer.

PLDs, the precursors of FPGAs, have actually been used to replace SSI in glue logic, yet they did not help much in prototyping. In addition, PLDs implement logic in AND/OR, and all the flip-flops are at the periphery of the devices. This type of logic arrangement restricts the designer and minimizes the flexibility.

FPGAs combine the versatility of gate arrays and the programmability of PLDs. Unlike gate arrays, they do not require custom fabrication and are obtained off the shelf as are SSI and MSI chips. Because FPGAs can be programmed on the fly, they are definitely more suitable for prototyping than SSI and MSI chips. For example, Xilinx devices have been used in prototyping Intel's P5 microprocessor [5]. The hardware emulation made it possible to simulate the microprocessor at reasonable speed; thus the development time of the microprocessor was reduced. Reprogrammable FPGAs also allow the design to be altered and the chip reconfigured quickly. This ease of reprogrammability facilitates design changes, which is usually a costly process.

However, FPGAs have their limitations. For the same design implemented with FPGAs and PLDs, it is more likely that the PLDs will operate faster than the FPGAs. PLD performance is independent of the logic implemented. But for FPGAs the circuit delay depends on the performance of the design implementation tools. The delay parameters can be extracted after placement and routing, typically a time-consuming process. Also, the mapping of the logic design into the FPGA's architecture requires sophisticated design implementation (CAD) tools than PLDs. Compared with traditional gate arrays, FPGAs are less dense and operate at a lower speed; however, the rapid advance in FPGA technology is quickly closing the gap between the two realizations. Cost factor is the next consideration.

1.4.2 Cost of Design with FPGAs

Cost is an important factor favoring FPGAs. ASICs cost consists of two main components: fixed cost and variable cost. The fixed cost is also called the nonrecurring engineering (NRE) cost. This is the initial cost that is needed to prepare masks, buy design tools, etc. For FPGAs, NRE consists of the development system and the platform on which it is used. The platform can vary from a personal computer to a more sophisticated workstation. NRE also includes the cost of the CAD tools (for design entry, simulation, and implementation) and the device programmer. The variable cost for FPGAs is that of the chip itself, while for gate arrays, it consists of the costs of testing and packaging the individual chips.

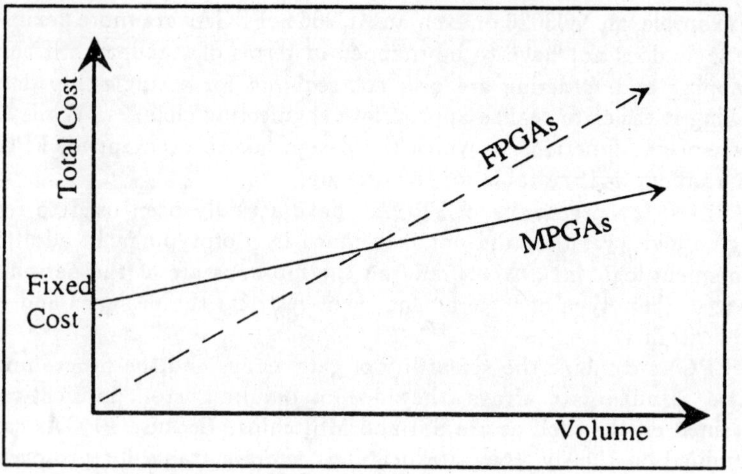

Figure 1.5: Cost Versus Design Volume.

Manufacturing testing cost for FPGAs is greatly reduced because test generation is done once for the unprogrammed chip. This will be discussed again in Chapter 8. At present, gate densities for FPGAs are lower, on the average, than are those of MPGAs. As the density is constantly increasing, FPGAs are becoming more competitive with gate arrays.

The curves in Fig. 1.5 show the change in cost as a function of the design volume needed. For gate arrays, there is a large fixed cost (at zero volume). The total cost then increases at a slower rate than the increase in the FPGAs cost. For FPGAs the cost is constant. Judging only from the cost of the product, FPGAs are more viable for lower production volume. It is interesting to realize that for most ASICs, the volume is usually not high, as illustrated by the distribution shown in Fig. 1.6.

In general, increased design flexibility comes at the cost of ease of implementing user logic, as Fig. 1.7 illustrates. Comparison of the different attributes of the different design implementations is given in Table 1.1 [11]. Since different ASIC devices have different advantages and disadvantages, the choice of a specific ASIC device largely depends on the user's application and needs.

1.5 The Design Cycle

The design process for using FPGAs generally requires nine steps:

1. Entering the design in the form of schematic, netlist, logic expressions, or hardware description languages
2. Simulating the design for functional verification

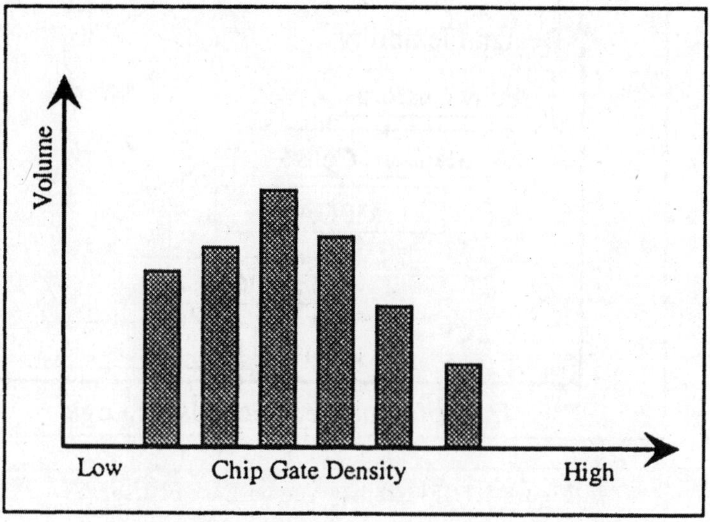

Figure 1.6: Volume Versus Circuit Size.

Characteristic	FPGA	Gate Arrays	Standard Cells	Full Custom
Design time	Short	Short	Short	Long
Fabrication	–	Short	Long	Long
Chip area	Very large	Large	Intermediate	Small
Cost	Very low	Low	Intermediate	High
Versatility	Very low	Low	Intermediate	High
Design cycle	Very short	Short	Intermediate	Long

Table 1.1: Comparison of Design Methodologies.

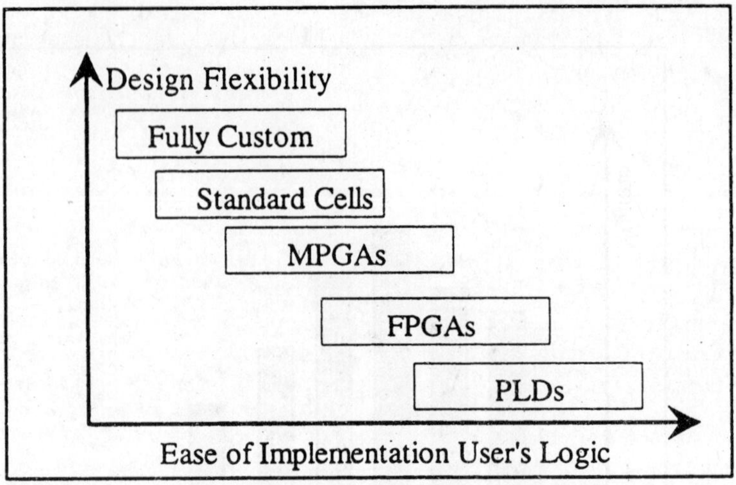

Figure 1.7: Flexibility Versus Ease of Use.

3. Mapping the design into the FPGA architecture
4. Placing and routing the FPGA design
5. Extracting delay parameters of the routed design
6. Resimulating for timing verification
7. Generating the FPGA device configuration format
8. Configuring or programming the device
9. Testing the product for undesirable functional behavior

Most FPGA vendors allow the entry of the design in a schematic form, though it is also possible to enter the design in Boolean expressions using, say, PALASM [10] or ABEL [4]. The most versatile design entry, however, is some hardware description languages. The most popular languages are `Verilog` HDL and VHDL. Commercially, the use of HDL has not been very widespread; but it is growing. Such languages require logic synthesis tools prior to mapping into FPGAs. In this book, we shall use **bdsyn** (an HDL translator developed at University of California, Berkeley [13]) to illustrate this approach of design entry.

Skipping the simulation steps enables the designer to obtain the end product faster, but this can undermine the product quality. It is a particularly costly omission for one-time programmable devices. Typically, the design goes through several iterations of simulation.

For every FPGA, the vendor provides design implementation tools to perform steps 3 through 8, and possibly step 6. The front-end design entry (schematic capture or other modes) and simulators may also be part of the tools. Most vendors configure their package with different front-end tools to allow the users more choice and flexibility.

Steps 3 and 4 involve several processes: logic minimization, technology mapping, placement, and routing. Technology mapping binds the technology-independent description of the circuits to the basic entities of the target technology. Placement allocates these entities to a specific physical block on the device. Then routing establishes the connections between the different blocks, and is usually done in two stages: *global routing* and *detailed routing.* All FPGA vendors have an automatic placement and routing tool. The placement and routing algorithm has much bearing on the performance of the design. Placement and routing are the topics of Chapter 7.

1.6 Verification and Testing

Testing is an important process in a product life cycle. It is performed at different stages - design, manufacturing, and use. At the design stage, testing is known as *design verification.* The most common means of verification is simulation. Simulation itself is a multilevel process, ranging from high level to low level, as we alluded to in the previous section. More information on simulation is given later in this chapter. Formal verification involves proving the properties of a design to guarantee correctness. Formal verification is still in the research and development stage.

Testing after manufacturing is usually known as *digital testing.* This type of testing detects not only failures due to manufacturing defects but also failures due to incorrect design. Often, simulation patterns developed for design verification are complemented with patterns that are generated manually or by an automatic test pattern generator (ATPG) to obtain a complete test set, a test that also verifies the functionality of its logic [1]. The patterns are applied to the circuit using automatic test equipment (ATE). During its lifetime, a product may be tested because it failed or because it is being serviced. The product may also be tested during normal operation. This is known as *online* or *concurrent testing.* Concurrent testing makes use of extra hardware constructs that have been designed exclusively for this purpose. To facilitate digital testing, testing strategies are now an integral part of the design. Designing with testing in mind is called *design for testability (DFT).*

In Chapter 8, the concepts of testing are presented, and we shall differentiate between testing the unprogrammed device and verification of the final design implementation. We shall also discuss types of faults specific to FPGAs and programmable devices in general.

1.7 Design Implementation (CAD) Tools

Computer-aided design tools have greatly facilitated the design implementation process. The tools have replaced many of the tedious tasks that a designer used to carry out – synthesis (logic minimization, technology mapping, state reduction, and state assignment), design entry, and design verification. Done manually, these tasks are tedious and time consuming and often error prone.

1.7.1 Simulation

Simulation is a process that imitates the functionality or behavior of the digital design on a computer. It is mainly used to identify possible design errors in the design or timing problems in a circuit. Simulation employs *models* that represent those system's attributes to be imitated; this may include behavioral or timing models. Simulation was originally used in lieu of *prototyping.* Now, simulation is often used to debug a prototype. FPGAs have also been used to prototype designs before their actual fabrication in other technologies, such as standard cells or gate arrays.

There are different aspects of digital circuits that need to be studied before implementation – functionality, timing, effect of certain parameters, etc. There are also a variety of simulation types that are dictated by circuit-level representations. For example, the functionality of a circuit may be simulated at the behavioral level, the gate or logic level, or the circuit level. In the case of FPGAs, the functional simulation can be performed on the behavioral level or the gate level. This simulation may be run with zero (no) delays, or one-unit delays. In the case of timing simulation, several approaches can be taken. Timing can be checked with nominal delays for the technology or with worst case scenarios. More important, timing can be verified using the actual layout of the design on the FPGA. For this, *back annotation* is necessary. The actual delays of the placed and routed design can be extracted and used in timing simulation. Simulation will be discussed further and illustrated by examples in Chapter 8.

1.7.2 Synthesis

Simply put, *synthesis* is the translation of a design representation to a form that is amendable to minimal realization. For example, the translation of a `Verilog` behavioral model into a register transfer language (RTL) is known as *high-level synthesis.* Mapping the RTL representation into a gate-level representation is part of *logic synthesis.* Logic synthesis also involves other processes used in the design of digital circuits such as logic minimization or state reduction and state assignment.

For logic minimization, Karnaugh maps can be used to find a minimal cover for a Boolean expression with a small number of inputs. The Quine-McCluskey method is more amenable to programming on a computer to accelerate and facilitate two-level logic minimization. However, since it is based on exhaustive comparisons and merging, it demands a fair amount of computer resources and time. The same problem arises in conjunction with state assignment for large finite state machines. Given a set of Boolean expressions, the time required to find the minimal cover for logic circuits increases tremendously with the number of terms to be minimized. Problems with such complexity are known as *NP complete*. Several heuristics have been developed and employed to address the minimal covering and state assignment problems.

In this book, we shall use several logic synthesis tools such as **espresso**, **misII**, and **mustang**. These programs have been developed at the University of California, Berkeley, and can be obtained from the University Industrial Relation Office for a nominal fee [13]: **espresso** maps minterms into a simplified two-level AND/OR expression; **misII** assists in multilevel logic minimization of Boolean functions; **mustang** is used as an example for state assignments. All three synthesis tools will be discussed in Chapters 5 and 6.

1.8 Things to Come

This chapter provides the context in which logic design and implementation will be presented in this book. Most topics will be detailed in subsequent chapters, which will also include a review of traditional logic design principles as a transition to designing with FPGAs.

It is important to realize that as this book is being written, changes and improvements in the FPGA technologies are constantly occurring. For example, when we wrote the prospectus for this book two years ago, we were using Xilinx XC3000 and ACT1 devices [15], [2]. Now Xilinx series 4000 is extensively used, and ACT3 has been announced. New architectures are being proposed [9], and improvements in programming devices have been published [3], [14]. Also, many third-party simulators and design capture programs are being developed. To date, three FPGAs workshops have been held [6], [8]. Also, standards to compare different devices have been developed and announced recently [12].

Bibliography

[1] Abadir, M., J. Furgson, and T. Kirkland. "Logic design verification via test pattern generation," *IEEE Transactions on Computer-Aided Design of Integrated Circuits and Systems, Vol. CAD-7*, pp. 138–149, January 1988.

[2] Actel. *Act Family Field Programmable Gate Array Databook*, Actel Corporation, Santa Clara, CA, March 1991.

[3] Chen, K.-L., et al. "A sublithographic antifuse structure for field-programmable gate arrays application," *IEEE Electron Device Letters*, Vol. 13, no. 1, pp. 53–55, January, 1991.

[4] Data I/O. "Programmable logic: A basic guide for designers." Data I/O Corporation, Santa Clara, CA, 1983.

[5] "Intel delays the P5 launch," *Electronics Engineering Times*, p. 1, July 27, 1992.

[6] FPGA '92, First Int'l Workshop on FPGAs (Berkeley, California), February 1992.

[7] Hollis, E. E. *Design of VLSI Gate Array ICs.* Englewood Cliffs, NJ: Prentice Hall, 1987.

[8] Moore, W., and W. Luk, ed. *FPGAs*, edited from the Oxford 1991 First Int'l Workshop on Field-Programmable Logic and Applications, Abingdon, England, 1991.

[9] Muroga, H., et al. "A large scale FPGA with 10K core cells with CMOS 0.8 μm," *Proc. of the Custom Integrated Circuit Conf.*, Paper No. 6.4, 1991.

[10] Monolithic Memories. *PAL/PLE Device Programmable Logic Array Handbook.* Monolithic Memories, Santa Clara, CA, 1986.

[11] Newell, S. B., A. J. de Geus, and R. A. Rohrer. "Design automation for integrated circuits," *Science*, Vol. 220, no. 4596, pp. 465–474, April 23, 1983.

[12] PREP. Programmable Electronics Performance Corporation, Los Gatos, CA, 1993.

[13] University of California at Berkeley, Industrial Liaison Office of Electrical Engineering and Computer Science Department, California, 1992.

[14] Wang, S.-J., et al. "High performance metal-silicide antifuse," *IEEE Electron Device Letters*, Vol. 13, no. 9, pp. 471–472, 1992.

[15] Xilinx. *The Programmable Gate Array Data Book.* Xilinx Inc., San Jose, CA, 1992.

Chapter 2

Field-Programmable Gate Arrays

2.1 Introduction

There is a wide variety of field-programmable gate arrays (FPGAs) as shown in Table 2.1. Some of these devices are actually programmable logic devices (PLDs) with specific expansions that make them larger and more flexible than traditional PLDs. For each vendor listed in the first column, the range of capacities for different devices is given. The data have been taken from vendors' literature and should be interpreted accordingly [1] – [4], [6, 7, 14, 17]. The effective capacities are usually lower since high utilization of the logic modules would generally impair routability. The architecture is identified in the third column as gate array (or row-based FPGAs), matrix form, sea-of-gates, or programmable logic array (PLA). We use the term "matrix" to mean that the logic blocks are organized in a matrix form with horizontal and vertical routing channels. We refer to the AND/OR structure of the first programmable logic array as PLA. The next column is the logic unit in which the user logic is implemented. The last column lists the programming method.

This chapter will focus on only two high gate-count devices: Xilinx logic cell array (LCA) and Actel application configurable technology (ACT) gate array. The first is taken as an example of RAM-based devices, and the second is an example of an antifuse-based type. Both devices are fabricated in complementary metal-oxide semiconductor (CMOS) technology.

In describing the two products, we shall focus on four factors that can influence the design and the selection of the device for a given application:

- Architecture
- Gate density
- Routing resources
- Programming method

Product	Capacity	Architecture	Basic Cell	Programming Method
Actel	2,000–8,000	Gate array	MUX	Antifuse
Algotronix	5,000	Sea-of-gates	Functional	SRAM
Altera	1,000–5,000	Extended PLA	PLA	EPROM
Concurrent	3,000–5,000	Matrix	XOR, AND	SRAM
Crosspoint	5,000	Gate array	Transistors	Antifuse
Plessey	2,000–40,000	Sea-of-gates	NAND	SRAM
QuickLogic	1,200–1,800	Matrix	MUX	Antifuse
Xilinx	2,000–10,000	Matrix	RAM block	SRAM

Table 2.1: Examples of FPGAs.

2.2 The Xilinx Logic Cell Array

The Xilinx logic cell array family was introduced in 1983 [5, 10]. Since then the product has passed through three generations: series XC2000, XC3000, and more recently the XC4000 [18]. In this chapter we shall focus on the Xilinx 3000 series, XC3000. However, we shall also refer to the other series for comparison and highlighting the special features.

Feature	XC2000	XC3000	XC4000
Equivalent gates	1,200 – 1,800	2,000 – 9,000	2,000 – 10,000
Number of CLBs	64 – 100	64 – 320	64 – 400
Inputs per CLB	4	5	9
Flip-flops per CLB	1	2	2
I/O Blocks	58 – 74	64 – 144	64 – 240
Maximum RAM bits	0	0	2,000 – 20,000
Functions per CLB	1 of 4 variables	1 of 5 variables	1 of 5 variables
			1 of 4 variables
	2 of 3 variables	2 of 4 variables	1 of 9 variables

Table 2.2: Features of the Xilinx Devices (April 1993).

Table 2.2 summarizes the main features of the three generations of Xilinx devices. For entries where a range is given, the devices corresponding to the upper values may or may not be available. The reader should refer to the vendor literature. The number of equivalent gates capacity (two-input NANDs) serves as a guide for a designer to select the appropriate part type. The estimation of the configurable logic block (CLB) requirement to implement a given set of logic expressions will be discussed in Chapter 5. The LCA architecture includes a fixed array of CLBs, input/output Blocks (IOBs), as shown in Fig. 2.1. The blocks can be connected by horizontal and vertical wire segments

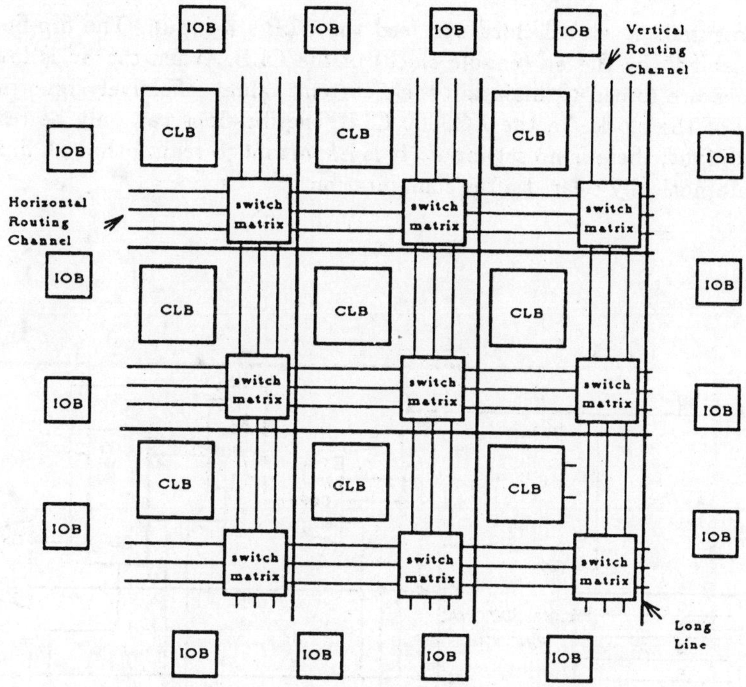

Figure 2.1: Xilinx: XC3000 LCA Architecture.

joined by the switch matrices. The programming elements inside the switch matrices establish the connections between the wire segments.

User logic is implemented by configuring the LCA's components. The Xilinx chip is reprogrammed since its programmability is based on the SRAM technology. The number of CLBs in an LCA ranges from 64 in the XC2064, the low end of the 2000 series, to 400 in the XC4010, the largest device presently available of the 4000 series (as of April 1993).

2.2.1 Configurable Logic Block

The CLB of the 3000 series consists of two flip-flops, a five-input look-up table, and several internal multiplexers, as shown in Fig. 2.2. The look-up table is a block of 32-bit RAM cells. The input variables (a,b,c,d,e) are applied to the address lines of the RAM cells. The look-up table can realize any single function of five variables. It is also possible to partition the look-up table into two 16-bit partitions to produce two function generators of four variables each. The two outputs of the CLB, x and y, may be connected to the outputs of the function generators F and G directly or via the flip-flops built in the CLB. Because the function generators are implemented using a RAM-based look-up table, propagation delay through the CLB is independent of the logic implemented. A global clock can feed the CLB's K input. The flip-flop can be "enabled" by the ec (enable clock) of the CLB. When the ec is low, the flip-flops are forced to maintain their current values, effectively ignoring the arrival of the clock. In the XC3000 CLB, the flip-flops can only be reset by the rd input; there is no set input. It is important to realize that all flip-flops are automatically cleared after configuration.

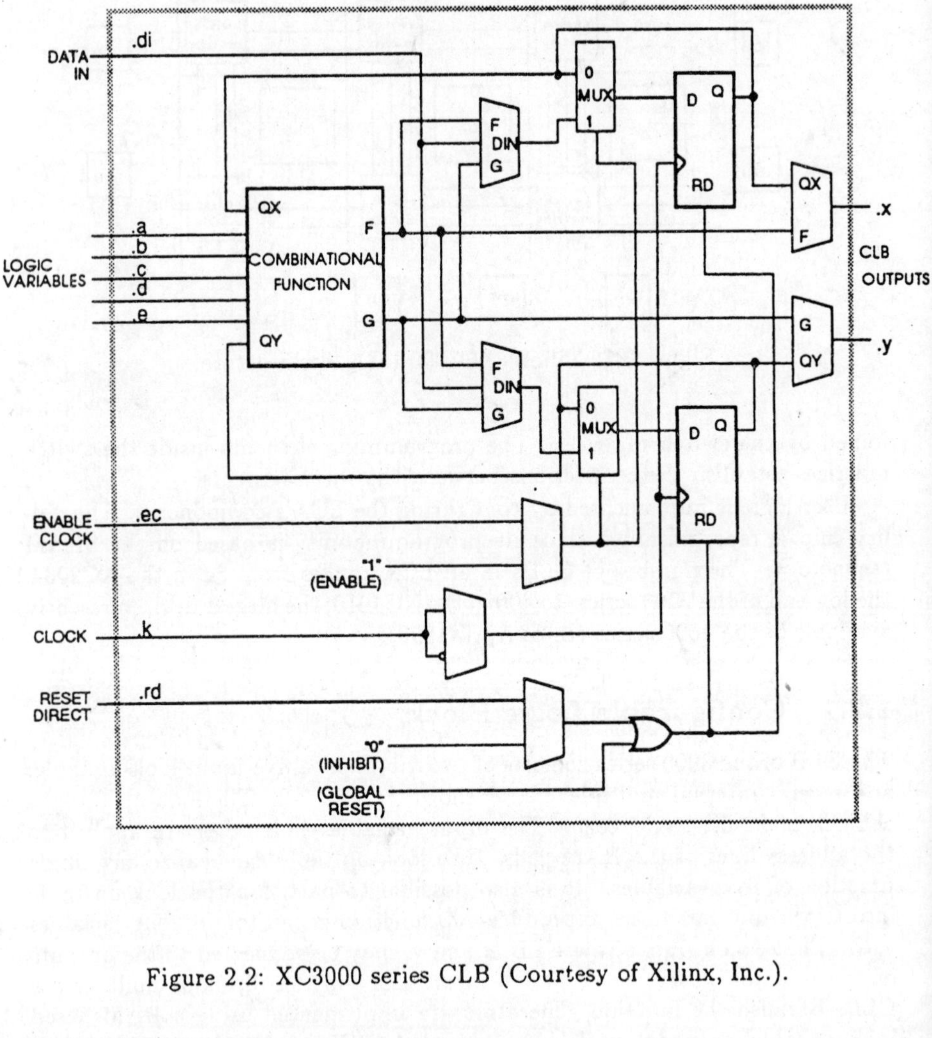

Figure 2.2: XC3000 series CLB (Courtesy of Xilinx, Inc.).

The CLB for the 4000 family is more versatile. Its block diagram is shown in Fig. 2.3. The 4000 CLB has three function generators, `F`, `G`, and `H`. `F` and `G` are four-input function generators, whereas `H` has three inputs. Both registered `XQ`, `YQ` and unregistered outputs `X`, `Y` are available at the output of the CLB. In the XC4000 CLB, the flip-flops can be both reset and set by the `S/R` input.

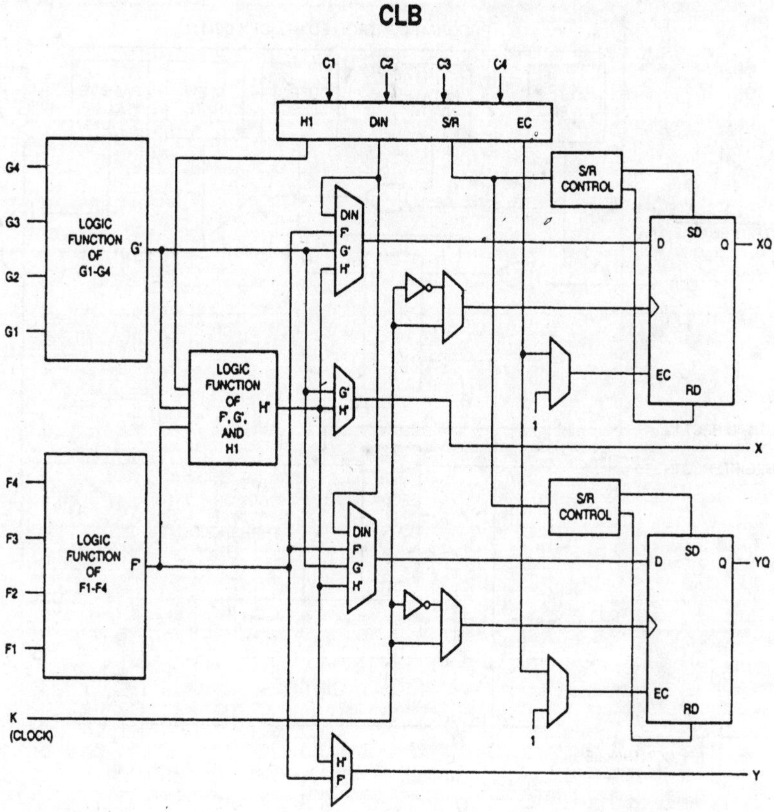

Figure 2.3: The XC4000 series CLB. Shown are the F and G function tables (computing any function of inputs F1–F4 and G1–G4, respectively), the H function table (computing any function of inputs F, G, and H1), various configurable multiplexers (denoted by trapezoids), and the two flip flops (Courtesy of Xilinx, Inc.).

2.2.2 The I/O Block

The XC3000 IOB, as shown in Fig. 2.4, provides an interface between the LCA and the external world. There is one IOB for every programmable pin

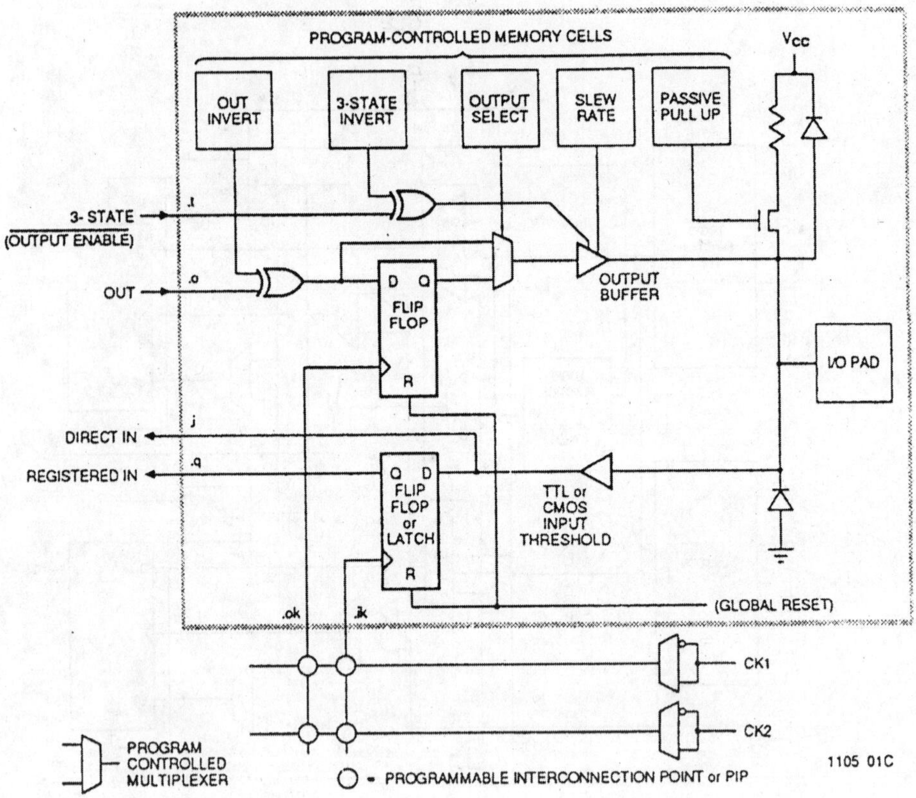

Figure 2.4: The I/O Block of XC3000 Devices (Courtesy of Xilinx, Inc.).

on the LCA package. The XC3090, for instance, provides 144 such I/O pins. Each IOB contains input and output buffers to facilitate compatibility with TTL and CMOS threshold levels. The IOB can serve as an input, output, or tristated bidirectional path. The input signal can be introduced to the internal logic directly or via a D-flip-flop (registered) or a transparent latch. Similarly, the output can optionally be registered or directly connected to the pad The output buffer is provided with skew control as well as tristate control. Generally, the number of I/O pads and IOBs increases with the logic capacity of the part type. There may be some unbonded pads that are inaccessible to the IOBs. Electrically, the XC3000 IOB can either sink or source 4 mA of current. External buffers are usually needed to interface the IOB to current demanding devices.

The 4000 family has an additional I/O feature, the *boundary scan* circuitry, which greatly facilitates board-level interconnect testing. The boundary scan will be discussed later in the chapter. The XC4000 IOB can either sink 12 mA or source 2 mA of current.

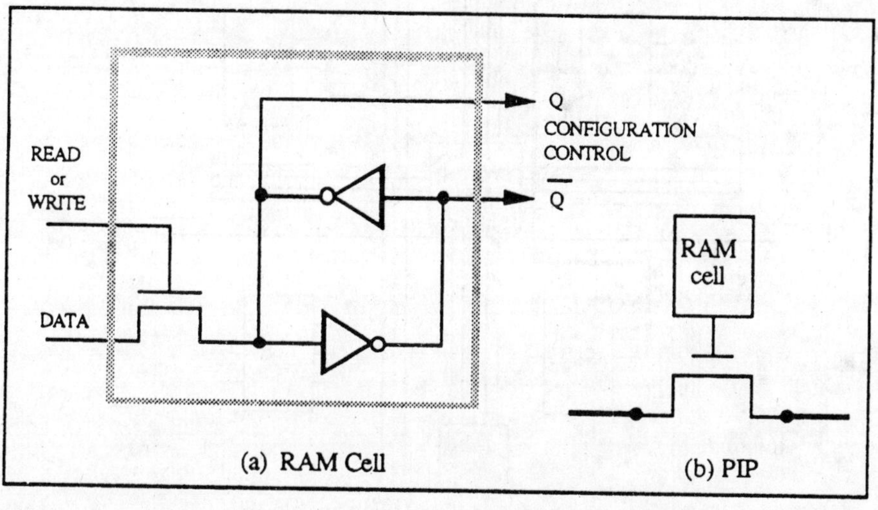

Figure 2.5: The Programmable Interconnect: RAM Cell and PIP (Courtesy of Xilinx, Inc.).

2.2.3 Programmable Interconnects

Implementing an entire design on the LCA requires interconnecting the various CLBs and IOBs. This is facilitated by the programmable interconnect resources, which consist of a grid of two layers of metal segments, *programmable interconnect points* (PIPs) and switch boxes. A PIP comprises a pass transistor that is controlled by a configurable RAM cell, as shown in Fig. 2.5. Dropping a "1" into the RAM cell establishes a connection between two points.

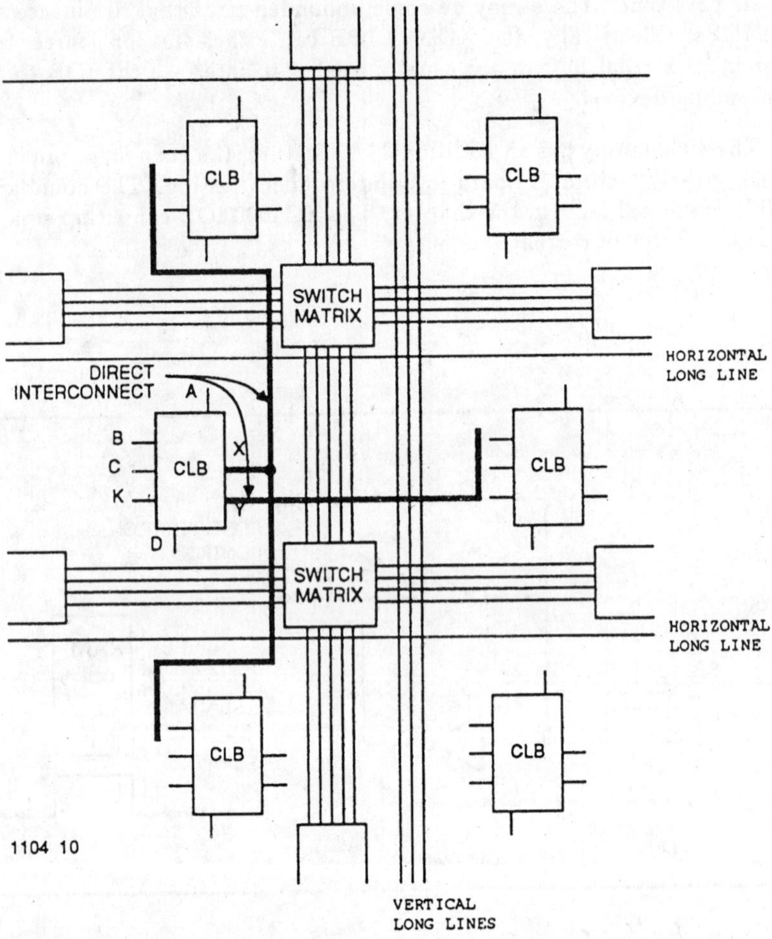

Figure 2.6: Xilinx: General Interconnects (Courtesy of Xilinx, Inc.).

Different routing resources are available for different circumstances. Vertical and horizontal long lines run the entire height and width of the interconnect area. They are used to carry high-fanout signals or signals that need to travel long distance with low skew. Direct connection is available between adjacent CLBs and IOBs for short-distance and low-delay communication. General-purpose interconnect is the most versatile of all. It consists of a network of switching matrices and wire segments to facilitate general network branching and routing.

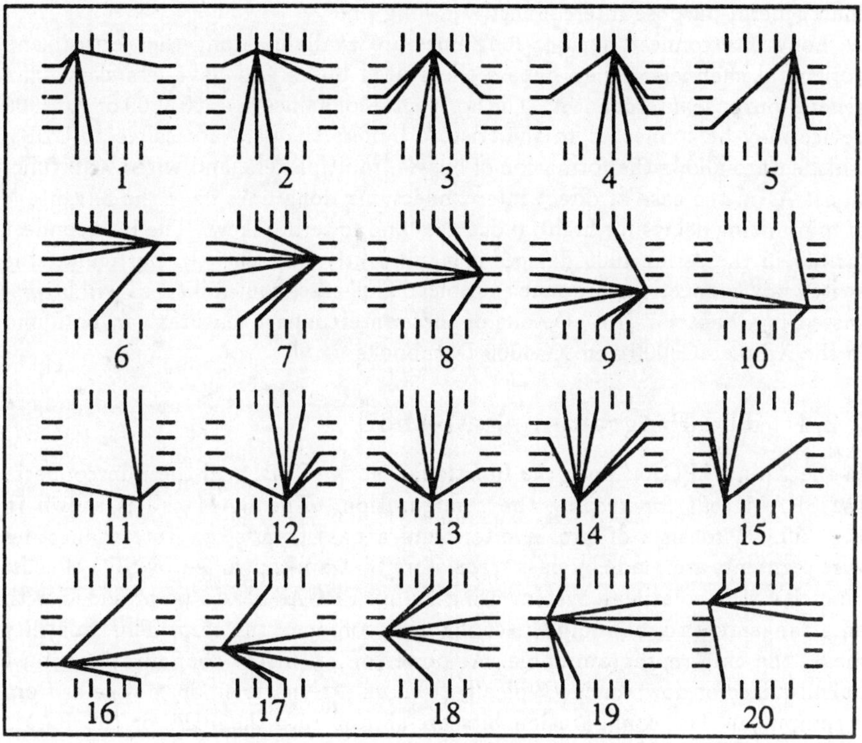

Figure 2.7: Possible Switch Box Connectivities (Courtesy of Xilinx, Inc.).

The general-purpose interconnects are to the Xilinx device what the routing channels are to gate arrays. In the XC3000 family, they consist of five horizontal and five vertical lines located between the rows and columns of CLBs. At the intersection of vertical and horizontal lines, there is a switch box, also called a *switching matrix*, as shown in Fig. 2.6. Twenty admissible interconnect options of a Xilinx XC3000 switch box are depicted in Fig. 2.7. Some signal-restoring buffers are available along some routing resources to restore signal levels due to degradation through the pass transistors inside the switch boxes.

Each XC2000 and XC3000 CLB can be directly connected to its neighboring blocks. The X-output of a CLB is directly connected to the B-input of the CLB immediately to its right and the C-input of the CLB to its left. The Y-output is connected to the A- and D-input of the CLB below and above, respectively.

Direct interconnect is also provided between the CLBs (along the logic cell boundary) and some IOBs. Direct interconnect typically has less routing delay than general-purpose interconnect and long lines.

Long interconnect lines or long lines are available along the vertical and horizontal channels. They bypass the switch boxes and have less skew than general-purpose interconnect. The horizontal long lines in a XC3000 or XC4000 device may be connected to the tristate buffers that are accessible to CLBs. This feature allows the formation of busses, multiplexers, and wired-AND functions. As in the case of direct interconnections, long lines have the advantage of minimizing delays for high-fanout nets and lowering skew. The interconnect facility in the series 4000 devices is significantly different. In particular, the switch box connections are much simpler. This eases routability, as will be discussed in Chapter 7 [15]. Details of these interconnect features can be found in the Xilinx XC3000 and XC4000 Databooks.

2.2.4 The Programming Method

In the Xilinx FPGAs, both the functional block (CLB) and the interconnects use SRAM cells for keeping the configuration. The RAM cell is shown in Fig. 2.5. It consists of two inverters and a pass transistor. Interconnecting wire segments are made by pass transistors that are controlled by SRAM cells. Since RAM is volatile, a system using Xilinx FPGAs has to be provided with an arrangement to maintain the configuration map of the chip. This volatility makes the chip reprogrammable. At power on, all RAM configurations must be initialized or downloaded typically from an external memory device or from a computer. The configuration bits are shifted into the PIPs or the RAMs inside the CLBs using a built-in shift register chain.

2.3 Advanced Features of the 4000 Series

There are several features in the 4000 series that resulted in an improvement in design flexibility and in device performance [18]:

- CLBs can be used as on-chip RAMs
- Fast-carry chain
- Boundary scan (JTAG) compatibility
- Wide decode logic
- More global clocks
- Faster placement and routing algorithms

- Scaled routing resources
  - Regular architecture
  - More PIPs

On-Chip RAM. The XC4000 CLB, which is shown in Fig. 2.3, consists of three function generators, F, G, and H. The F and G function generators can be used as a 16×1 RAM or together as a 32×1 RAM. The address of the RAMs is applied through the inputs labeled F1-F4 and G1-G4. The C1-C3 inputs are used as write-enable and data-ins, respectively.

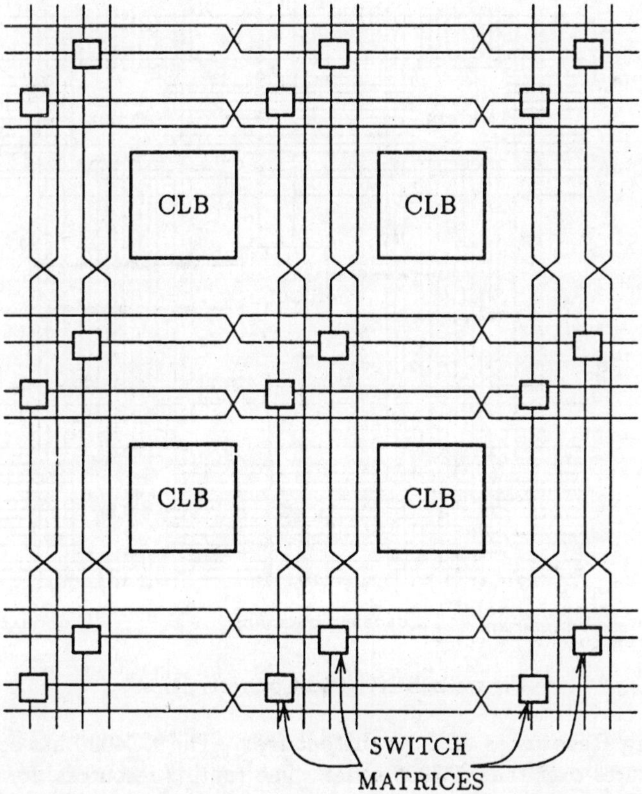

Figure 2.8: XC4000 Routing Resources Showing the Double-Length Lines (Courtesy of Xilinx, Inc.).

Fast Arithmetic. In the XC3000 devices, a full adder can be implemented in one CLB with the sum and the carry obtained at the outputs X and Y. The XC4000 can be configured to add two 2-bit words using the fast-carry chain. Blocks F and G generate the two sums, and the carry can be forwarded directly to the next CLB above or below. The implementation of this logic is accomplished only through *hard macros*. A hard macro is a preplaced and prerouted logic module using the low-level graphic design editor XACT. Direct control of this feature by the user at the schematic level will be available in the near future.

Boundary Scan. Boundary scan is a design for test (DFT) technique in which the I/O cells are configured as a shift register around the periphery of the chip for testing purposes [12]. Boundary scan operations are governed by an on-chip test access port (TAP) controller that works in accordance to the IEEE 1149.1 standard [11]. More details will be given in Chapter 8, where verification and testing are discussed.

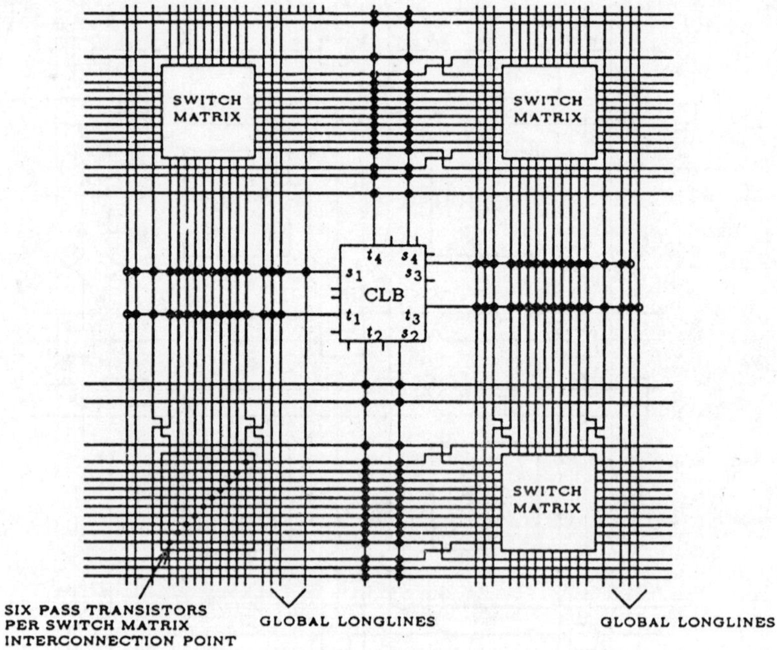

Figure 2.9: XC4000 Connectivities.

Routing Resources and Architecture. The XC4000 has several improved features over the XC3000 series. The routing resources, for example, form a totally symmetric array of potential connections as depicted in Fig. 2.8. This symmetry facilitates the routing by allowing the use of routing techniques developed for traditional ASICs. These resources include:

- More globally distributed signals; each XC4000 device has eight global buffers to support distribution of global signals.
- Double-length wires that consist of track segments, each track spanning two CLB lengths. Thus there are fewer PIPs per line as compared with the XC3000 devices.

- Single-length wires intersecting through switch boxes, which allows every horizontal wire to connect to one vertical wire. In addition, the switch boxes have fewer and simpler connections, as shown in Fig. 2.9.
- Connecting CLBs outputs and inputs to a high percentage of the wiring segments, as shown in Fig. 2.9.

In addition, routing resources are scaled across the different part types in the XC4000 family, unlike the XC3000 family. For instance, an XC4010 device has roughly twice the routing tracks available in an XC4003 device.

2.4 The Actel ACT

In this section, the Actel FPGAs are contrasted sharply to the Xilinx FPGAs, both in architectures and programming methods. At present, there is more than one vendor that produces antifuse-based FPGAs, as indicated in Table 2.1. Actel is the representative in the antifuse- and segmented channel-based FPGAs. Actel's ACT gate array was introduced in 1988 [8], and it has two generations: ACT1 and ACT2. The third-generation ACT3 has already been announced. The chip consists of rows of logic modules, peripheral circuits, and routing channels as illustrated in Fig. 2.10.

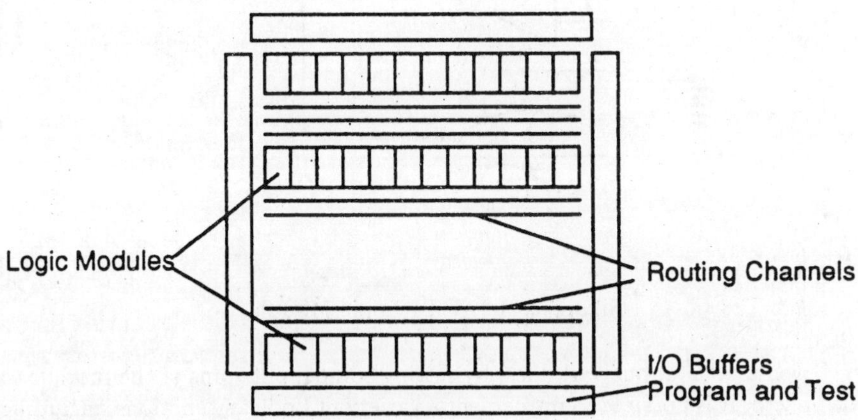

Figure 2.10: Actel FPGA Architecture.

User logic is implemented using Actel's proprietary programmable low-impedance circuit element (PLICE) antifuse technology. The programmability of the PLICE is irreversible, making the Actel FPGA a one-time-programmable device.

2.4.1 The Logic Module

The building block of the ACT architecture is the logic module, shown in Fig. 2.11. The number of modules per chip ranges from 295 in the smallest (ACT1010) to 1,400 in the largest device (ACT1260). The ACT1 logic module is an eight-input and one-output cell. Each logic module consists of three 2-to-1 multiplexers and one two-input OR gate.

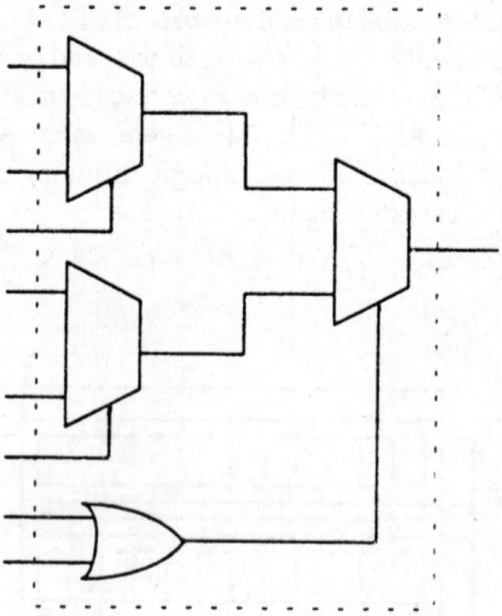

Figure 2.11: ACT1 Logic Module.

Given a set of input variables, by setting some inputs pins of the module to one or zero, it is possible to implement a variety of Boolean functions, including all two-variable Boolean functions and some three- and four-variable Boolean functions. For more complex functions, more than one module is used. The *Actel Databook* [1] shows a whole library of functions. There are no dedicated flip-flops in the ACT1. All required registers are built from combinations of logic modules.

A transparent latch can be configured in one module, while a flip-flop needs two modules. The ACT2, however, has flip-flops that are categorized as se-

	ACT1010	ACT1020	ACT1130	ACT1260
	1988	1988	1989	1989
Number of gates	1,200	2,000	3,000	6,000
CMOS process	2.5	2.5	2.5	1.25
Logic modules	295	546	823	1,400
Possible flip-flops	147	273	411	1,400
Possible latches	295	546	823	1,400
PLICE antifuses	112,000	186,000	400,00	666,000
Number of I/Os	57	69	112	160

Table 2.3: ACT Array Characteristics.

quential (S-block) and combinational (C-block) devices. Each S-block includes an edge-triggered D-flip-flop, but it takes two C-blocks to form one flip-flop. The ACT array family characteristics are summarized in Table 2.3.

2.4.2 Peripheral Circuits

Peripheral circuits in the ACT gate array consist of I/O buffers, testability circuits, and diagnostic probe circuits. The I/O buffers allow each pin to be configured as an input, output, tristate, or bidirectional buffer. The testability and diagnostic probe circuits in conjunction with two diagnostic probe pins allow the user to observe any internal signals *after programming* the chip.

2.4.3 Routing Channels

The routing resources in an Actel FPGA are shown in Fig. 2.12. Routing channels contain several segmented metal tracks with PLICE antifuses. They provide all interconnections between logic modules and I/O modules. Contrary to the Xilinx LCA architecture, the channels in this device run along one dimension of the chip. Similar to traditional gate arrays, the cross-directional routing are fed through over the rows of blocks are included in the routing resources. All eight inputs of the logic module are connected to the routing channels, four to each surrounding channel. The output of the module is connected to both channels through a vertical track that spans two channels above and two below.

2.4.4 Actel's PLICE

The programmable low impedance circuit element antifuse is a nonvolatile, two-terminal interconnect element. It occupies the area of a via (connection between two metal layers) in a mask-programmable gate array and forms a conducting path when configured.

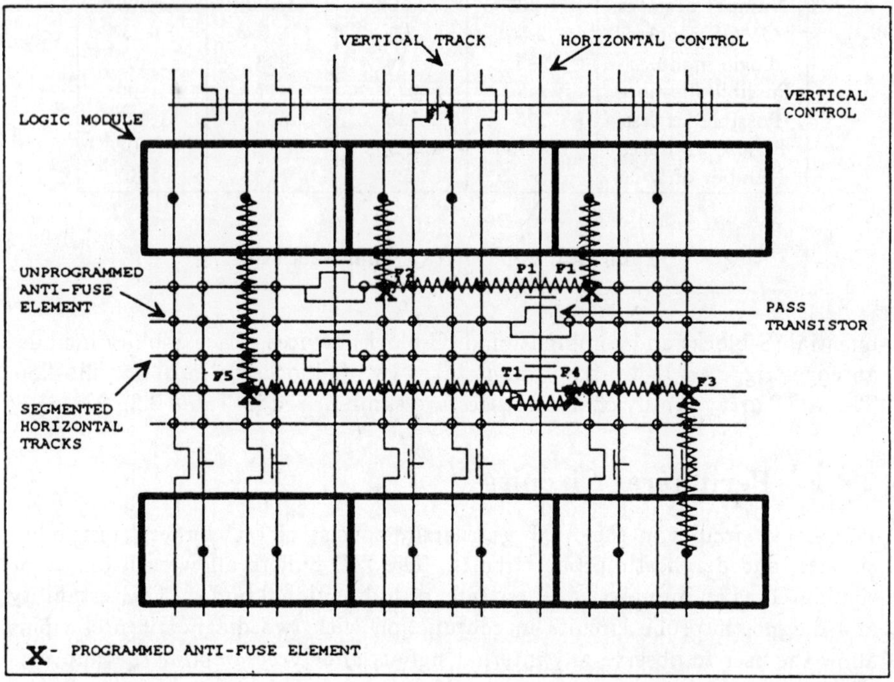

Figure 2.12: ACT1 Routing Resources.

The PLICE consists of a dielectric layer between two conducting materials, polysilicon and diffusion, as shown in Fig. 2.13. Each PLICE is at the intersection of a vertical and a horizontal path. When the appropriate voltages are applied across the two paths, a link is formed connecting its two terminals by a very low resistance. The PLICE's ON-resistance varies between 300 Ω and 500 Ω, while its off-resistance is larger than 100 MΩ [9]. The programming requires an 18 volt source that supplies 5 mA through the devices. Other FPGAs, such as QuickLogic and Crosspoint, use metal-to-metal antifuses with lower ON-resistance of 30 Ω [4]. QuickLogic ViaLink uses a thin layer of amorphous silicon between the two metal layers. The antifuse is equivalent to a via in a two-metal CMOS process.

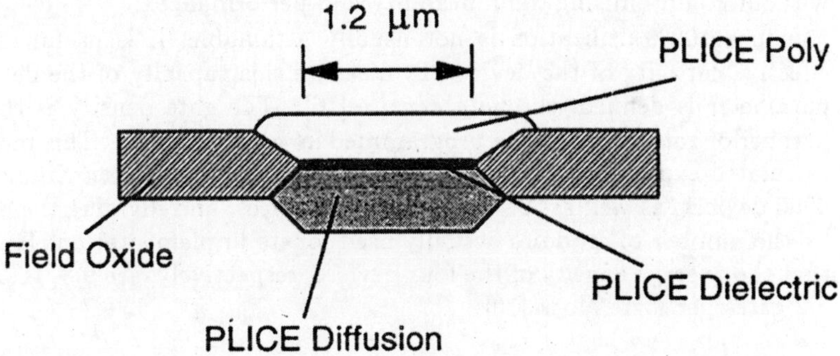

Figure 2.13: Actel's PLICE.

2.5 Technology Trends

The wide variety of FPGAs offers different solutions to the problem of achieving highly integrated, fast, and flexible devices. All these devices offer levels of architectural capabilities in programmable devices that have never been available before. Most vendors supply development systems that significantly ease design and shorten the development cycle. More important, these devices afford an environment in which testing issues can be addressed with less concern over design methodology trade-off.

2.5.1 Device Capacity, Utilization, and Gate Density

ASIC capacities have been typically measured by the equivalent number of NAND gates on the IC. The number of gates is calculated by counting the number of transistors on the IC and dividing this number by the number of transistors required to implement a NAND gate for the particular technology. The highest gate count devices available at present are the Xilinx's XC4010 with 10,000 (claimed) gates and Actel's ACT2 1260 with 8,000 gates. This does not imply that the XC4010 can implement larger designs than Actel's ACT2 1260. Measuring ASIC capacities using the number of equivalent gates can be misleading since it does not relate a meaningful standard that applies across all ASIC types [13].

Another measure that is more appropriate to FPGAs is *utilization*. It is defined as the number of logic blocks used for actual realization of the design. To use a device to its full capacity is not recommended; that is, to reach

100% utilization, since it would impair the routability of the design. The architectural challenge in designing an FPGA is to achieve a high utilization without compromising the routability and performance.

Since 100% utilization is not usually attainable, it is useful to find the effective capacity of the device. To measure this capacity of the device a new parameter is defined, the ***gate density*** [16]. The gate density is the average number of gates that can be programmed in a logic module. This measure was calculated experimentally by implementing several designs on Xilinx 3000 and 4000 devices, as well as ACT1 and ACT2 devices, and dividing the design size by the number of modules actually used for its implementation. It was found that the average density of the four devices, respectively, are 9.3, 16.5, 3.5, and 6.2 gates per logic block [16].

Feature	SRAM	EPROM	Antifuse
Technology	CMOS submicron	Standard double polysilicon	New process polysilicon and diffusion or amorphous
Programming Method	Shift register	FAMOS UV erasure	Break down
Area	Very large	Large	Small
Resistance	$\simeq 2K\Omega$	$\simeq 2K\Omega$	Actel $\simeq 500\Omega$
Capacitance	$\simeq 50fF$	$\simeq 15pF$	$\simeq 5fF$

Table 2.4: Comparing Programming Elements.

2.5.2 Programming Method

Sizable differences among the FPGAs relate to their interconnect programming methods and logic cell structure. The programming method affects both the interconnect delays and the type of application. Unlike the ACT antifuse, the Xilinx LCA's configuration is volatile, and all memory is lost when the device is powered down. For this reason, the LCA must be used in conjunction with an external memory device. In addition, since the LCA is not programmed directly, no specialized programmers need to be purchased and the LCA can be reconfigured on the fly if the application so demands (*dynamic reconfiguration*).

The other parameter affected by the programming method is the area of the logic cell. An SRAM logic cell may consist of five transistors, thus occupying a larger area than a multiplexer-based logic cell such as the Actel FPGA, which

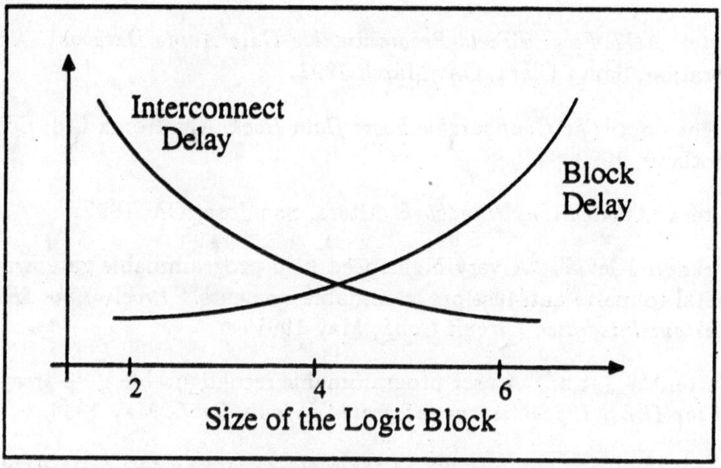

Figure 2.14: Effect of the Block and Interconnect Delays.

use antifuses for configuring the cells. Table 2.4 gives a comparison among the three main programming devices: SRAM, antifuse, and EPROM.

In addition to the size, the electrical characteristics of the programming element affect routing and the performance of the device. However, the effect of these parameters cannot be evaluated independently of the logic block size and its fanin. As the block size increases, more functionality can fit within a block, and the number of connectivities is decreased. However, as the fanin increases, the delay through the block will also increase. The conflicting variations of the logic block delay and the interconnect delays are illustrated in Fig. 2.14, where both types of delays are plotted versus the fanin to the block. This trend has been confirmed with experimental results [15].

2.5.3 Summary

This chapter is not meant to be an exhaustive description of FPGAs. Our purpose is to describe the salient features of these devices that are pertinent to the design process. The two selected types of devices, Xilinx and Actel, are chosen because they differ in architecture and in programming method. The emphasis was more on Xilinx devices because it is the type used as the implementation example throughout the book.

Bibliography

[1] Actel. *ACT Family Field-Programmable Gate Array Databook.* Actel Corporation, Santa Clara, CA, March 1991.

[2] Algotronix. *The Configurable Logic Data Book.* Algotronix Ltd, Edinburgh, Scotland, 1990.

[3] Altera. *Applications Handbook.* Altera, San Jose, CA, 1992.

[4] Birkner, J. et al. "A very high speed field-programmable gate array using metal to metal anti-fuse programmable elements." *Proc. of the IEEE 1991 Custom Integrated Circuit Conf.*, May 1991.

[5] Carter, W., et al. "A user programmable reconfigurable gate array." *Proc. of the IEEE 1986 Custom Integrated Circuit Conf.*, May 1986.

[6] Concurrent Logic. *CFA6000 Series Field-Programmable Gate Arrays.* Concurrent Logic, Sunnyvale, CA, 1992.

[7] Crosspoint. *CLi 6000 Series Field Programmable-Gate Arrays.* Crosspoint Logic, San Jose, CA, 1992.

[8] El-Ayat, et al. "A CMOS electrically configurable gate array," *IEEE Journal of Solid State Circuits,* Vol. 24, no. 3, pp. 752–762, June 1989.

[9] Hamdy, et al. "Dielectric based antifuse for logic and memory IC." *International Electron Device Meeting*, pp. 786–788, May 1988.

[10] Hsieh, H. C., et al. "Third generation architecture boosts speed and development of FPGA." *Proc. of the IEEE 1990 Custom Integrated Circuit Conf.*, pp. 31.2.1–31.2.7, May 1990.

[11] IEEE. "Standard test access port and boundary-scan architecture." Report sponsored by the Test Technology Technical Committee of the IEEE Computer Society, Document P1149.1/D5 (draft), June 1989.

[12] Maunder, C., and F. Beenker. "Boundary scan framework for structured design for test." *Proc. of the International Test Conf.*, pp. 714–723, 1987.

[13] Osann, B., and A. El-Gamal. "Comparing ASIC capacities with gate array benchmarks," *Electronic Design*, October 13, 1988.

[14] Plessey. *CMOS Semi-custom CLA 60000, ASIC Handbook.* Plessey Semiconductor, 1991.

[15] Rose, J., and S. Brown. "The effect of switch box flexibility on routability of FPGAs." *Proceedings of the IEEE 1990 Custom Integrated Circuit Conf.*, pp. 27.5.1–27.5.4, May 1990.

[16] Small, B., A. Bodmer, and S. Mourad. "Evaluating FPGAs: Capacity and utilization." *Proceedings of the Phoenix Conf. on Comp. and Comm.*, (Phoenix, Arizona), pp. 134–140, March 1993.

[17] Xilinx. *The Programmable Gate Array Data Book.* Xilinx, Inc., San Jose, CA, 1992.

[18] Xilinx. *The XC4000 Data Book.* Xilinx, Inc., San Jose, CA, 1992.

Chapter 3

Design Flow

3.1 Introduction

This chapter provides an overview of the steps to be taken from entering a design to realizing a design in a single field-programmable gate array (FPGA). Designing with multiple FPGAs is more involved, and the discussion will be relegated to Chapter 9. The process involves

1. Design specification
2. Design simulation/analysis
3. Design entry, schematic drawings
4. Technology dependent/independent minimization
5. Logic partitioning, technology mapping
6. Physical partitioning
7. Placement and routing
8. Back annotation and timing verification or simulation
9. Realization by programming an FPGA

In very brief terms, the specification of a design is presented in either abstract terms or in formal methods, followed by analyzing the feasibility of the implementation of the design by high-level simulation. Then the implementation of the design is entered (or expressed) in the description that is admissable by the FPGA vendor's tools for instance, in terms of gates/macro cells that are in the vendor's schematic capture macro cell library. The more technology-dependent processes start from here. Technology mapping involves transforming a technology-independent description of a circuit into a representation that matches the architectural/technological constraints of the target technology, which will be FPGAs. At this point, only the contents of the basic cells of the target technology are defined; *where* the cells should be placed and *how* to connect them remain to be determined. This is the task of an automatic placement and routing tool, for example, apr. The implementation of the design is completed at this point, but it takes one final step to *program*

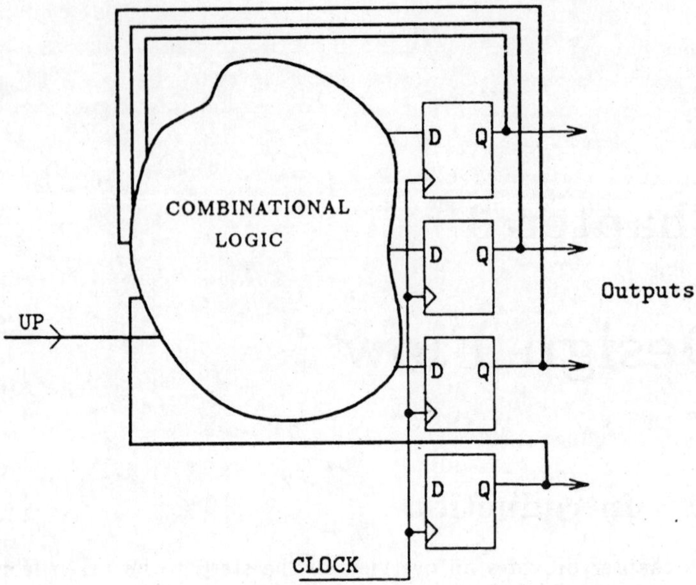

Figure 3.1: A 4-Bit Up/Down Counter.

the device (FPGA) to realize the design. Notice that the level of abstraction gradually reduces as we approach the realization of the design. We shall detail the design flow process in subsequent sections.

3.2 Design Specification and Minimization

This is the "thinking" and "writing" processes at the abstract/high level. This means that the language of thinking and writing should be as close as to the "human" level as possible. Humans are good at thinking in "abstract" terms. For example, when we say "counter" or "decrementer," perhaps the most familiar picture that would come to our minds is of the "+" and "–" symbols, but not chains of flip-flops and combinational logic. The following example illustrates a high-level description in the form of **bdsyn** language [10].[1] This will be our "canonical example" and will also be referred to in subsequent chapters.

Figure 3.1 illustrates a 4-bit up/down counter controlled by the **up** bit. Figure 3.2 describes the *combinational logic* of the 4-bit up/down counter in the bdsyn language. The high-level operators "+" and "–" are abstract in the sense that they match our "common sense" notion of increment and decre-

[1]The author finds this language appealing for classroom purposes because it has the flavor and merits of high-level description languages but without the large overhead involved in learning a commercial high-level description language.

ment, but not in terms of flip-flops and logic gates. The combinational logic of the counter has been specified in a RTL-like language **bdsyn**. The **bdsyn** is a hardware description translator which takes as input a textual description of a block of combinational logic and produces a collection of Boolean functions that implement the described combinational logic. **Bdsyn** outputs *berkeley logic interchange format* file (**blif**) that can be used by a logic minimization program misII for further minimization.

```
!
! Combinational Logic of a 4-Bit Up/Down Counter
!
MODEL counter
      D<3:0> =                        ! outputs
      Q<3:0>, up;                     ! inputs

    BEHAVIOR;

    PORT
    Q<3:0>, up input,
    D<3:0> output;

          ROUTINE counter;

                 IF (up EQL 1) THEN
                              D = Q + 1
                 ELSE
                              D = Q - 1;

          ENDROUTINE counter;

    ENDBEHAVIOR;
ENDMODEL counter;
```

Figure 3.2: High-Level Description of the Combinational Logic of a 4-Bit Up/Down Counter.

MisII is a multiple-level combinational logic minimization program [2], and is part of the Berkeley logic synthesis system (superseded by SIS as of 1993). The figure of merit of the minimization is the *literal count*. The literal count of a variable in an expression is the number of times the variable or its negation appears in the right-hand side of the expression. The literal count of an expression is the sum of the literal count of all its variables. For example, the literal count of the following expression

$$a = bc' + dc$$

is four.

At the present time, if you have little idea of what **misII** is, think of it as a tool that can identify *covers* in a Karnaugh map. This tool can handle a large number of Boolean variables, and can implement a Boolean function in multiple levels of logic. **MisII** attempts to minimize the number of literals appearing in a set of logic expressions; **misII** can output logic expressions in Boolean equation (**eqn**) format. For example, the unminimized **blif** netlist format file generated by **bdsyn** from the 4-bit up/down counter description is shown in Fig. 3.3.

A more readable description of the combinational logic of the up/down counter is shown in Fig. 3.4 as unminimized Boolean equations. Figure 3.5 shows a set of minimized expressions by **misII**.

There are other "high-level" constructs in the **bdsyn** language that could not be described in complete detail here. The best source is [10]. Just to illustrate, Fig. 3.6 shows a **bdsyn** description of a 3-bit up/down counter controlled by the up bit. This description uses the high-level operator **SELECT**. If the up signal is high, the operator **SELECT** decodes the Q input and provides the D output, which is one more than the input.

Bdsyn is limited to combinational logic description. There are separate tools to deal with the specification of the sequential portion of the design. For example, **mustang** is a state assignment program [4]. **Mustang** outputs a *programmable logic array* format (**pla**) that can be used by **espresso** and **misII** to reduce the combinational logic of the finite state machine. All these programs are distributed by the Department of Electrical Engineering, University of California at Berkeley. They are available to other universities at minimal cost. Please write to

Industrial Liaison Program Software Distribution Department of Electrical Engineering and Computer Sciences 457 Cory Hall, University of California, Berkeley Berkeley, CA 94720

3.3 Design Entry

There are several ways to enter a design to implement with XILINX FPGAs. We shall describe two "commercial" paths to enter and implement a design, followed by the description of an "academic" path.

3.3.1 Commercial Tools

The commercial paths depend largely on schematic capture tools or schematic editors for design entry, as illustrated in Fig. 3.7. A schematic editor is just

```
.model counter
.inputs Q3 Q2 Q1 Q0 up
.outputs D3 D2 D1 D0
.names Q0 counter/$PLUS_5_1/sum0.1.1
0 1
.names Q1 Q0 counter/$PLUS_5_1/sum1.1.1
01 1
10 1
.names Q1 Q0 counter/$PLUS_5_1/c0.1.3
11 1
.names Q2 counter/$PLUS_5_1/c0.1.3 counter/$PLUS_5_1/sum2.1.1
01 1
10 1
.names Q2 counter/$PLUS_5_1/c0.1.3 counter/$PLUS_5_1/c0.1.4
11 1
.names Q3 counter/$PLUS_5_1/c0.1.4 counter/$PLUS_5_1/sum3.1.1
01 1
10 1
.names Q0 counter/$MINUS_4_1/sum0.2.1
0 1
.names Q1 Q0 counter/$MINUS_4_1/sum1.2.1
11 1
00 1
.names Q1 Q0 counter/$MINUS_4_1/c0.2.3
1- 1
-1 1
.names Q2 counter/$MINUS_4_1/c0.2.3 counter/$MINUS_4_1/sum2.2.1
11 1
00 1
.names Q2 counter/$MINUS_4_1/c0.2.3 counter/$MINUS_4_1/c0.2.4
1- 1
-1 1
.names Q3 counter/$MINUS_4_1/c0.2.4 counter/$MINUS_4_1/sum3.2.1
11 1
00 1
.names up counter/$MINUS_4_1/sum0.2.1 counter/$PLUS_5_1/sum0.1.1 D0
01- 1
1-1 1
.names up counter/$MINUS_4_1/sum1.2.1 counter/$PLUS_5_1/sum1.1.1 D1
01- 1
1-1 1
.names up counter/$MINUS_4_1/sum2.2.1 counter/$PLUS_5_1/sum2.1.1 D2
01- 1
1-1 1
.names up counter/$MINUS_4_1/sum3.2.1 counter/$PLUS_5_1/sum3.1.1 D<3>
01- 1
1-1 1
.end
```

Figure 3.3: Unoptimized blif Netlist Format Representation of the 4-Bit Up/Down Counter.

```
INORDER = Q3 Q2 Q1 Q0 up;
OUTORDER = D3 D2 D1 D0;
  [0]  = Q0' Q1' Q2' Q3' up'
  [1]  = Q0 Q1 Q2 Q3' up
  [2]  = Q0 Q1 Q2' up
  [3]  = Q0' Q1' Q2' up'
  [4]  = Q2' Q3 up
  [5]  = Q1' Q2 Q3
  [6]  = Q0 Q3 up'
  [7]  = Q0 Q2 up'
  [8]  = Q1' Q2 up
  [9]  = Q0 Q1 up'
  [10] = Q0 Q1' up
  [11] = Q0' Q1 Q3
  [12] = Q0' Q1 Q2
  [13] = Q0' Q1' up'
  [14] = Q0' Q1 up
  [15] = Q0'
  [16] = [0]' [11]' [1]' [4]' [5]' [6]'
  {D3} = [16]'
  [18] = [12]' [2]' [3]' [7]' [8]'
  {D2} = [18]'
  [20] = [10]' [13]' [14]' [9]'
  {D1} = [20]'
  [22] = [15]'
  {D0} = [22]'
```

Figure 3.4: Unoptimized Logic Expressions of the 4-Bit Up/Down Counter.

```
INORDER = Q3 Q2 Q1 Q0 up;
OUTORDER = D3 D2 D1 D0;
  D3 = Q3' Q2' Q1' Q0' up' + Q3' Q2 Q1 Q0 up + Q3 Q0 up'
       + Q3 Q1 up' + Q3 Q2 up' + Q3 Q0' up + Q3 Q1' up
       + Q3 Q2' up;
  D2 = Q2' Q1' Q0' up' + Q2' Q1 Q0 up + Q2 Q0 up' +
       Q2 Q1 up' + Q2 Q0' up + Q2 Q1' up;
  D1 = Q1' Q0' up' + Q1 Q0 up' + Q1 Q0' up + Q1' Q0 up;
  D0 = Q0';
```

Figure 3.5: `MisII`-Minimized Logic Expressions of the 4-Bit Up/Down Counter.

```
!
! 3-bit incrementer, decrementer
! inputs: Q<2:0>, up
! up = 1 means increment, up = 0 means decrement
! outputs: D<2:0>
!
MODEL counter D<2:0> = Q<2:0>, up;

ROUTINE main;
   IF (up EQL 1) THEN
      BEGIN
         SELECT Q<2:0> FROM
            [0] : D<2:0> = 1;
            [1] : D<2:0> = 2;
            [2] : D<2:0> = 3;
            [3] : D<2:0> = 4;
            [4] : D<2:0> = 5;
            [5] : D<2:0> = 6;
            [6] : D<2:0> = 7;
            [7] : D<2:0> = 0;
         ENDSELECT;
     END
   ELSE
      BEGIN
         SELECT Q<2:0> FROM
            [0] : D<2:0> = 7;
            [1] : D<2:0> = 0;
            [2] : D<2:0> = 1;
            [3] : D<2:0> = 2;
            [4] : D<2:0> = 3;
            [5] : D<2:0> = 4;
            [6] : D<2:0> = 5;
            [7] : D<2:0> = 6;
        ENDSELECT;
      END;
 ENDROUTINE;
ENDMODEL;
```

Figure 3.6: High-Level Description of the Combinational Logic of a 3-Bit Up/Down Counter.

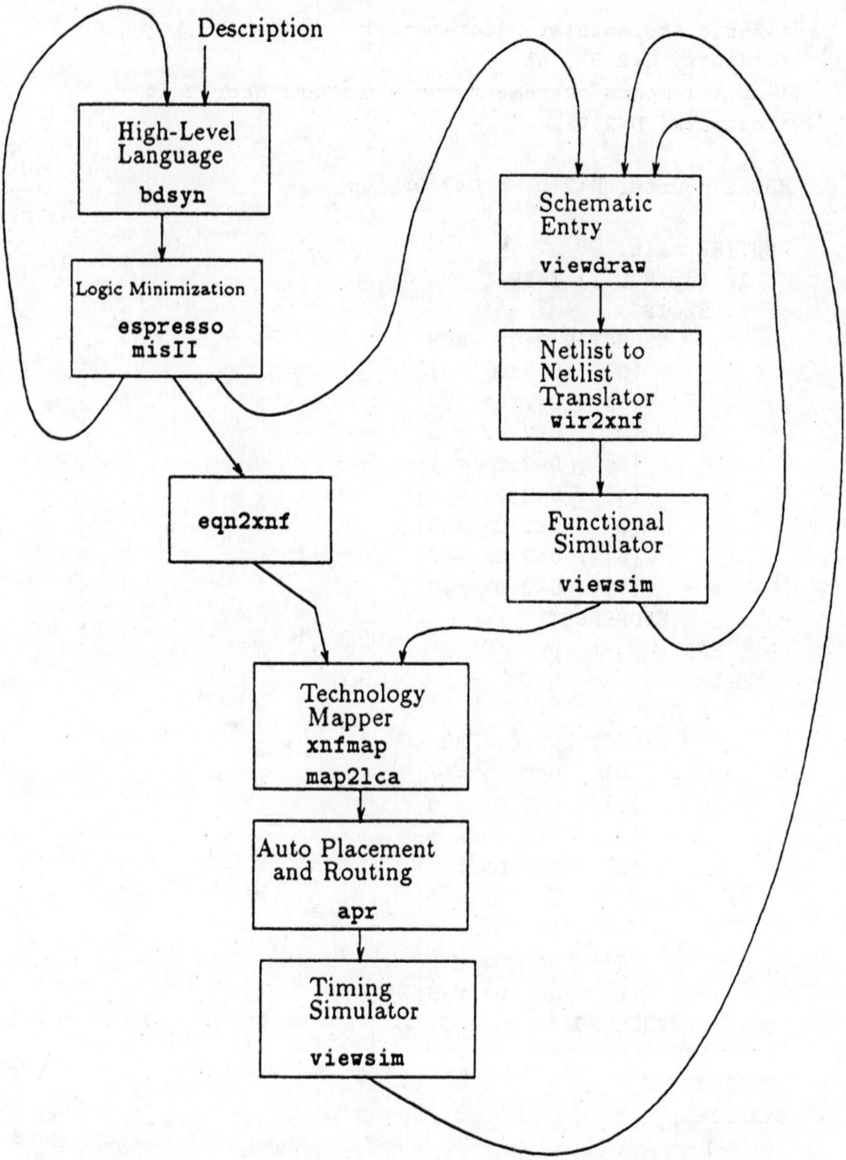

Figure 3.7: Design Flow Using a Hybrid of University and Commercial Tools.

a drawing tool that enables a designer to specify the *netlist* of a design by connecting the components. The components can come from a supplied library or can be made up by the user. We have used two industrial schematic editors `futurenet` (or `dash-lca`), which is a product of DATA I/O, and `viewdraw`, which is a product of Viewlogic to enter gate-level designs. There is a XILINX parts (macro/cell) library in either of the schematic editors [1]. Typically, the library contains basic logic gates, I/O pads, buffers, clock buffers, oscillators or pulse generators, latches, flip-flops, decoders, multiplexers, registers, counters, comparators, shifters, arithmetic functions, memories, and special functions. Some primitive library parts are illustrated in Figs. 3.8 and 3.9. Please consult your schematic-editor interface and macro library for details.

There are also special symbols for controlling and guiding the technology mapper, placement and routing during the implementation phase of the design. A good example is the CLBMAP in the XILINX 3000 macro library and FMAP (or GMAP) in the XILINX 4000 macro library. For instance, the CLBMAP symbol identifies the portion of the logic in a design that be implemented in a single CLB. Special functions in the XC4000 library include a boundary scan symbol (BNDSCAN) that is used to access the boundary scan logic in the XILINX XC4000 devices. All the XC3000 macros are *soft* macros, meaning that the placement and routing of the components are left to the mapper, and placement and routing tools. There are both *soft* and *hard* macros in the XC4000 macro library. The components in the hard macro library are premapped, prerouted, and preplaced. This enables mapping, placement and routing under the designer's control, instead of at the mercy of the tools. At the present time, the hard macros and `xblox` (see next paragraph) are the only means available to use the dedicated carry logic in a XC4000 device. Ideally, hard macros provide guaranteed routing, placement and routing of a mapped design, thereby guaranteeing the performance. The current (as of April 1993) limitation is that only mapping and placement are guaranteed but not the routing.

The traditional way of entering a design using a schematic editor is enhanced by the introduction of block/module-oriented entry tools such as `xblox`. `Xblox` enables a design to be entered at the "block" level, a level higher than gate level. In `xblox`, the width of data paths can be parameterized; therefore, a designer is relieved from entering the components along the data path one by one. This is illustrated in Fig. 3.10 which shows the schematic drawing of a Fibonacci sequence generator. The width of the data path can be altered by the parameter BOUNDS. `Xblox` also features optimization of data paths to some degree.

You are *discouraged* to enter your design *entirely* at the gate level. There are a number of facilities in the commercial schematic editors that enable a netlist description saved in a file to be imported and incorporated as part of the schematic drawing. For example, in both `viewdraw` and `futurenet`, the facility is the "FILE" attribute of a component/block. The advantage of this method of entry is that the results from a high-level description compiled

Figure 3.8: **Xnfwirec** XC3000 Gate Library.

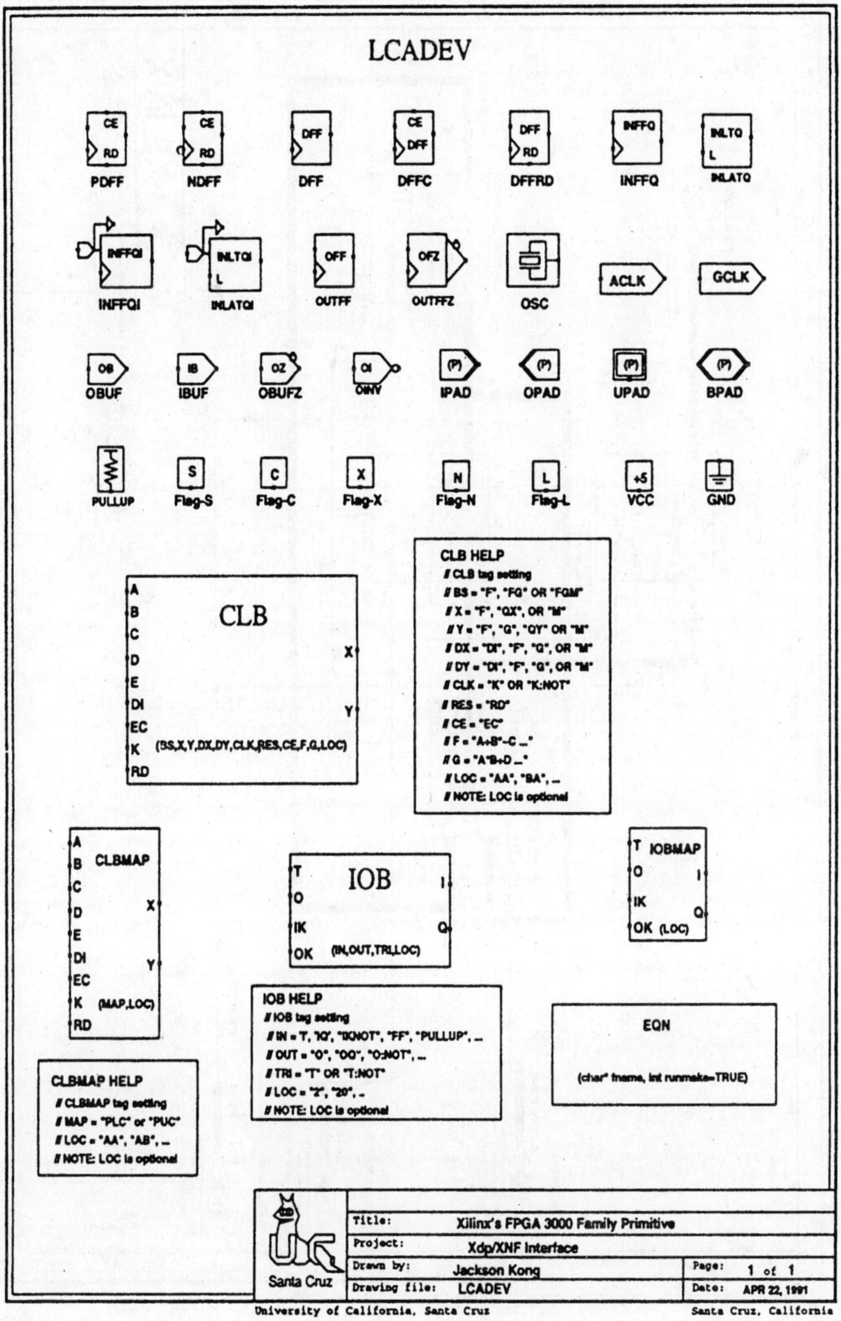

Figure 3.9: `Xnfwirec` XC3000 Device Library.

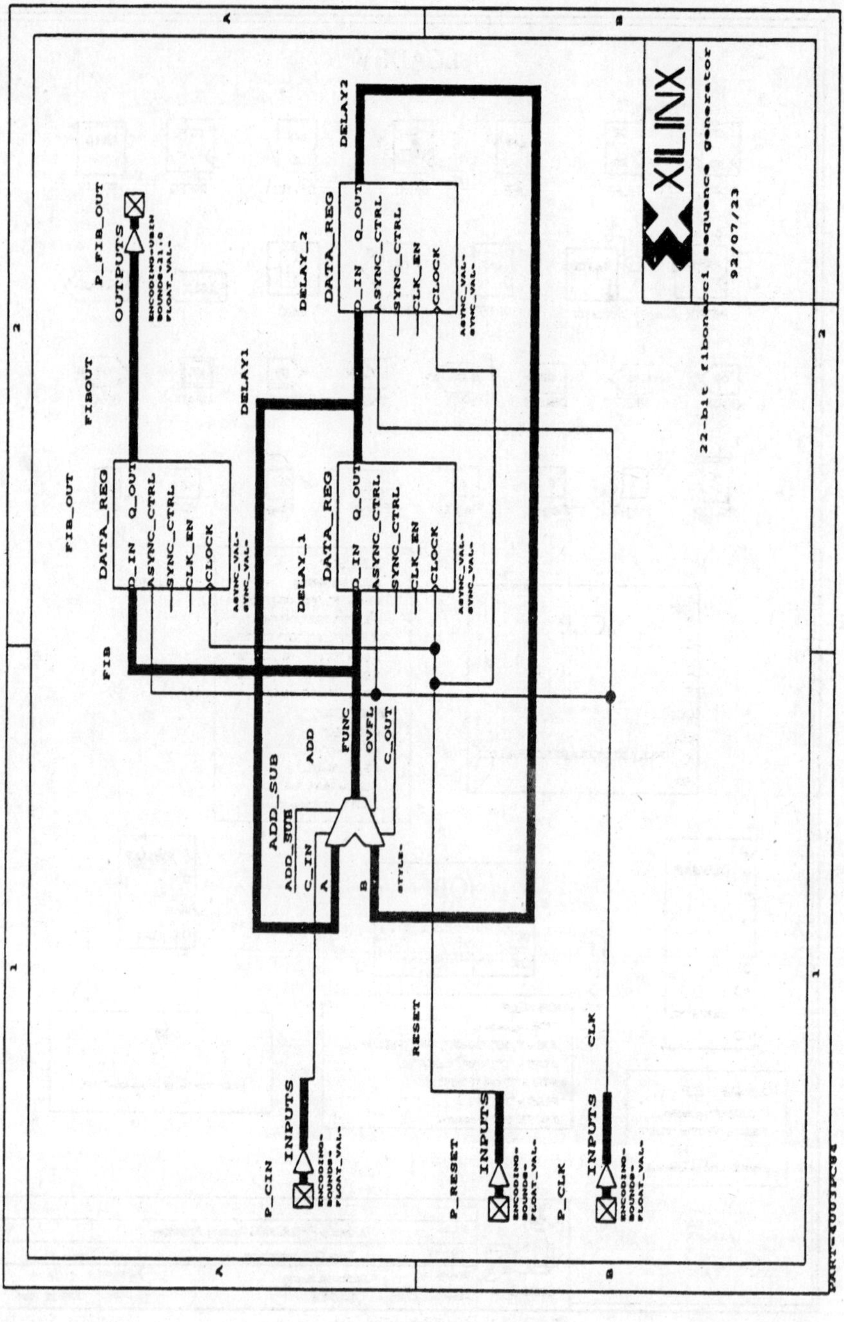

Figure 3.10: `Xblox` Example: A Schematic Drawing of a Fibonacci Sequence Generator (Drawn by Dr. Steven Kelem: One of the Developers of `xblox`).

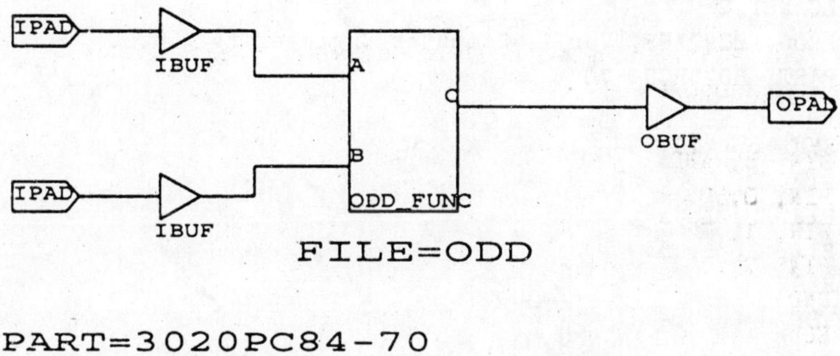

Figure 3.11: A Viewlogic Schematic Drawing **top** Using Functional Module (to Incorporate Boolean Equation Results from the Synthesis Tools).

down to gate level can be imported directly into a schematic drawing.[2] This is a hybrid approach that delegates the design using high-level tools and relegates the technology-dependent details to the schematic drawing tools. This is the approach that we shall advocate in this book.

Suppose that you have a top-level schematic diagram that defines the pin-in, pin-out, input pads, output pads, input buffers, and output buffers as shown in Fig. 3.11. The functionality of your design can be described in a separate **XNF** (Xilinx Netlist Format) file. Basically you need to specify the structure (topology) of your design with the schematic editor. The functionality of your design may be described in a high-level language. The description can be subsequently minimized by **misII**, which produces Boolean equations. From the **eqn** file generated by **misII**, you can convert it to the **XNF** format using a utility **eqn2xnf** developed at University of California Santa Cruz, which converts an equation netlist to an **XNF** netlist. You will need to use **xnfmerge** to merge/link all the **XNF** files together. **Xnfmerge** works like a linker that provides a flattened description of a hierarchical design.

Example 1: We create in **viewdraw** a schematic drawing of a two-input AND functional block **ODD_FUNC** as a *module*. Let us call this drawing **top** for future reference. The content of the functional block originates from an equation file **odd.eqn**:

```
INORDER = a b;
OUTORDER = c;
c = a*b;
```

[2]You need to flatten the design into a single **XNF** file using **xnfmerge** for logic simulation.

We translate the **eqn** file format to the **XNF** file format using **eqn2xnf**. The content of the **XNF** file **odd.xnf** is

```
LCANET,2
PROG, EQN2XNF, Ver 1, Computer Engineering @ UCSC
PART, 3020PC84-70

SYM, c, AND
PIN, O, O, c
PIN, 1, I, a
PIN, 2, I, b
END
EOF
```

After specifying, entering, or drawing the design, we shall start the implementation of the design from here. The gate-level netlist (**dwg** format in **futurenet**) (**wir** format in **viewdraw**) then needs to be converted to **XNF** format. This amounts to translating the netlist representation of a drawing tool to a more manufacturer-specific netlist representation. Then, the **XNF** netlist would be processed by a technology mapper, which converts a design presented by the netlist in a relatively *technology-independent* form and embedding it into a netlist representation that is specific to the architecture of the FPGA devices.

For example, in the XILINX XC3000 design flow, a 5-input AND gate that has been drawn using a schematic tool will be converted, by the technology mapper, into a 5-input lookup table with its content configured as an AND gate. After technology mapping, a partitioner generates an initial placement that determines roughly where the CLBs would be placed within the FPGA die. This process attempts to extract some global information about the natural structure of a mapped (but not yet placed) design. This is usually performed by a block partitioning program that exercises some form of partitioning algorithm on the design. Well-known partitioning algorithms are the Kernighan-Lee "mincut" algorithm [8], and the Fiduccia-Mattheyses "FM" algorithm [7]. These algorithms aim at locating clusters in a design and keeping the elements inside a cluster sufficiently "close" together to minimize some cost metric such as total wire lengths, or the number of interconnections in between the clusters.

The tool in the XC3000 family to accomplish the initial-partitioning task is **map2lca**. The result of this program is a partitioned or initially placed design represented in logic cell array (**LCA**) format. Notice that it is only initially placed; the placement and routing program **apr** is the authority to determine the final placement using a simulated annealing based algorithm.

In the XILINX XC4000 series tools, the design flow process is almost identical, but the technology mapper, block partitioner, place and route program are integrated into a single tool called **ppr**. Also, the placement algorithm in **ppr** is partitioning based instead of simulated annealing based as in **apr**.

Example 2: Suppose that you have an equation file name **odd.eqn**, and a

VIEWLOGIC schematic file **top** as your top-level drawing that uses the equations in odd.eqn. You need to convert the **eqn** format to **XNF** format first. Use the translator **eqn2xnf** to convert from the **eqn** format to the **XNF** format. Let us call the output file **odd.xnf**.

Here are some of the steps involved in using the basic design implementation tools:

```
%1>    eqn2xnf odd
%2>    wir2xnf -P3020PC84-70 top
%3>    xnfmerge top flat
%4>    xnfmap flat
%5>    map2lca flat
%6>    apr flat flat_route
%7>    xdelay flat_route
%8>    makebits flat_route
%9>    xchecker flat_route
```

The first two steps translate the netlists of the design to **XNF** format. The third step combines the two separate **XNF** netlists, **top.xnf** and **odd.xnf**, into a single file **flat.xnf**. The fourth step is technology mapping. The fifth step does mincut-based partitioning to obtain an initial placement. The sixth step involves detail placement and routing. The seventh step estimates the performance of the design using a static timing analyzer. The eighth step generates the configuration bitstream, and the final step downloads the bitstream to program the FPGA device.

For design verification and debugging, in particular using **workview**, here are some of the steps involved in using the back annotation and simulation tools:

```
%1>    lca2xnf flat_route  flatsim
%2>    bax flatsim flat -o flatsim
%3>    xnf2wir flatsim
%4>    vsm flatsim
%5>    workview
```

The first step translates the placed and routed design from **LCA** format to **XNF** format file **flatsim**. The second step uses **bax** to recover some of the original signal names that might have been eliminated by the technology mapper. The third step translates the back-annotated design file from **XNF** format to the VIEWLOGIC **WIR** format. The fourth step prepares the **WIR** file for simulation using **viewsim** in the **workview** package. The final step evokes **workview** for timing simulation.

3.3.2 University Tools

All the related university tools are well integrated in this path, which is illustrated in Fig. 3.12. The university tools are inside the dashed line. The process starts by using the **xdp/wireC** schematic editor to enter a gate-level design. There are XILINX parts library (both XC3000 and XC4000) available in **wireC** [5, 9]. The next two paragraphs are a brief description of **wireC** by one of its originators.

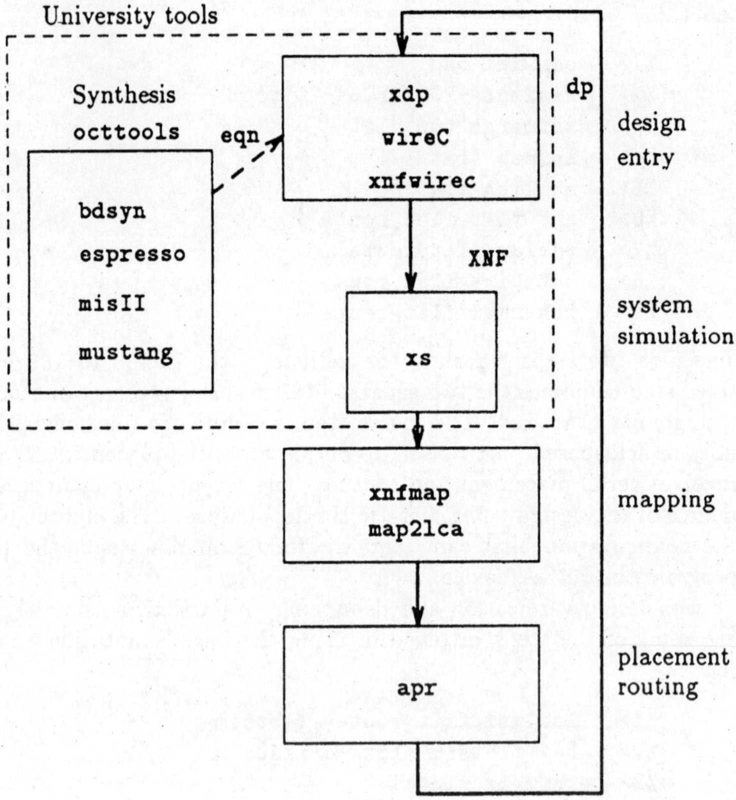

Figure 3.12: Design Flow Using University Schematic Drawing and Synthesis Tools.

WireC is a graphical specification language that combines schematics with procedural constructs for describing complex microelectronic systems. WireC allows the designer to choose the appropriate representation, either graphical or procedural, at a fine-grained level depending on the characteristics of the circuit being designed. Drawing traditional schematic symbols and their interconnections provides fast intuitive interaction with a circuit design, while procedural constructs give the power and flexibility to describe circuit structures algorithmically and allow single descriptions to represent whole families of devices.

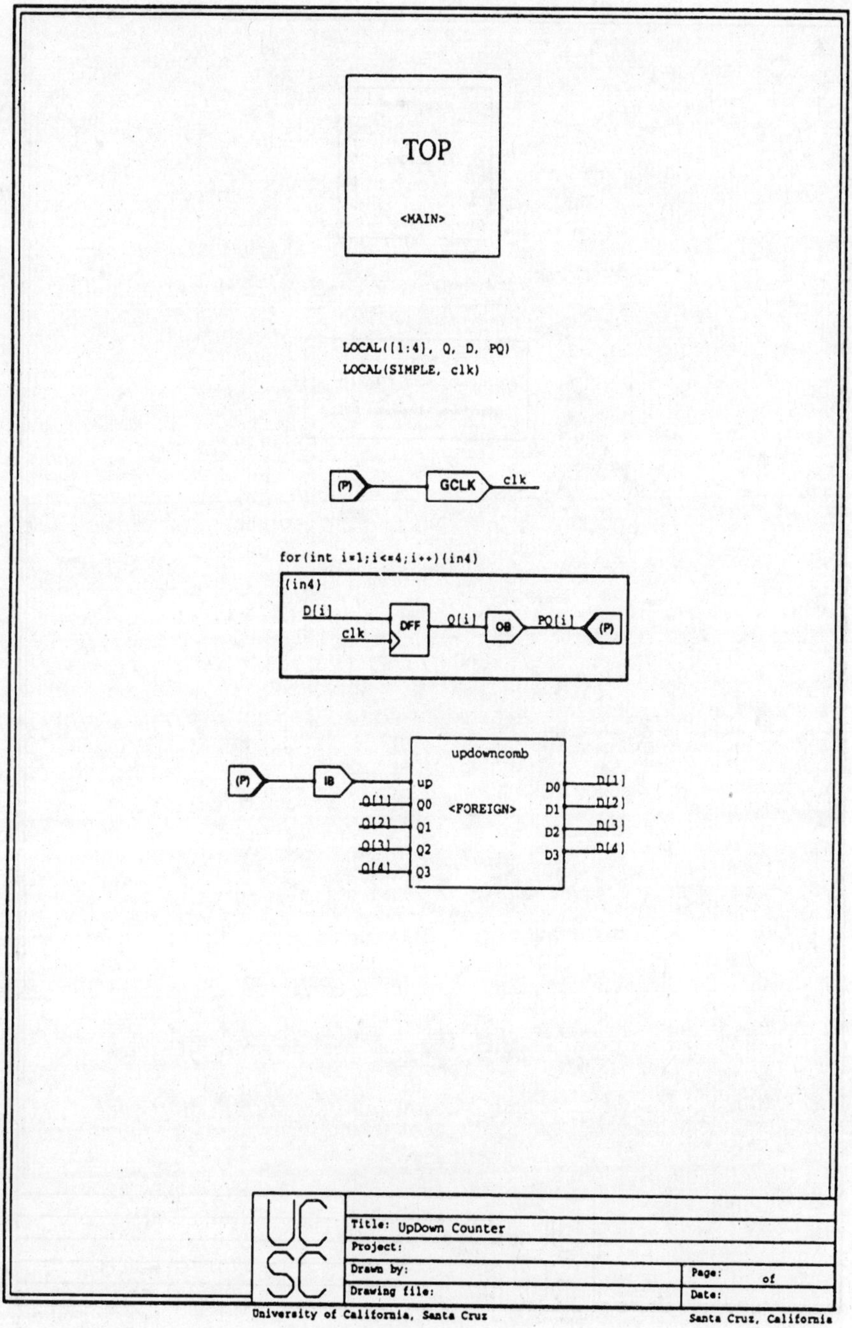

Figure 3.13: Top-level Drawing for an Up/Down Counter (Sheet 1 of 2).

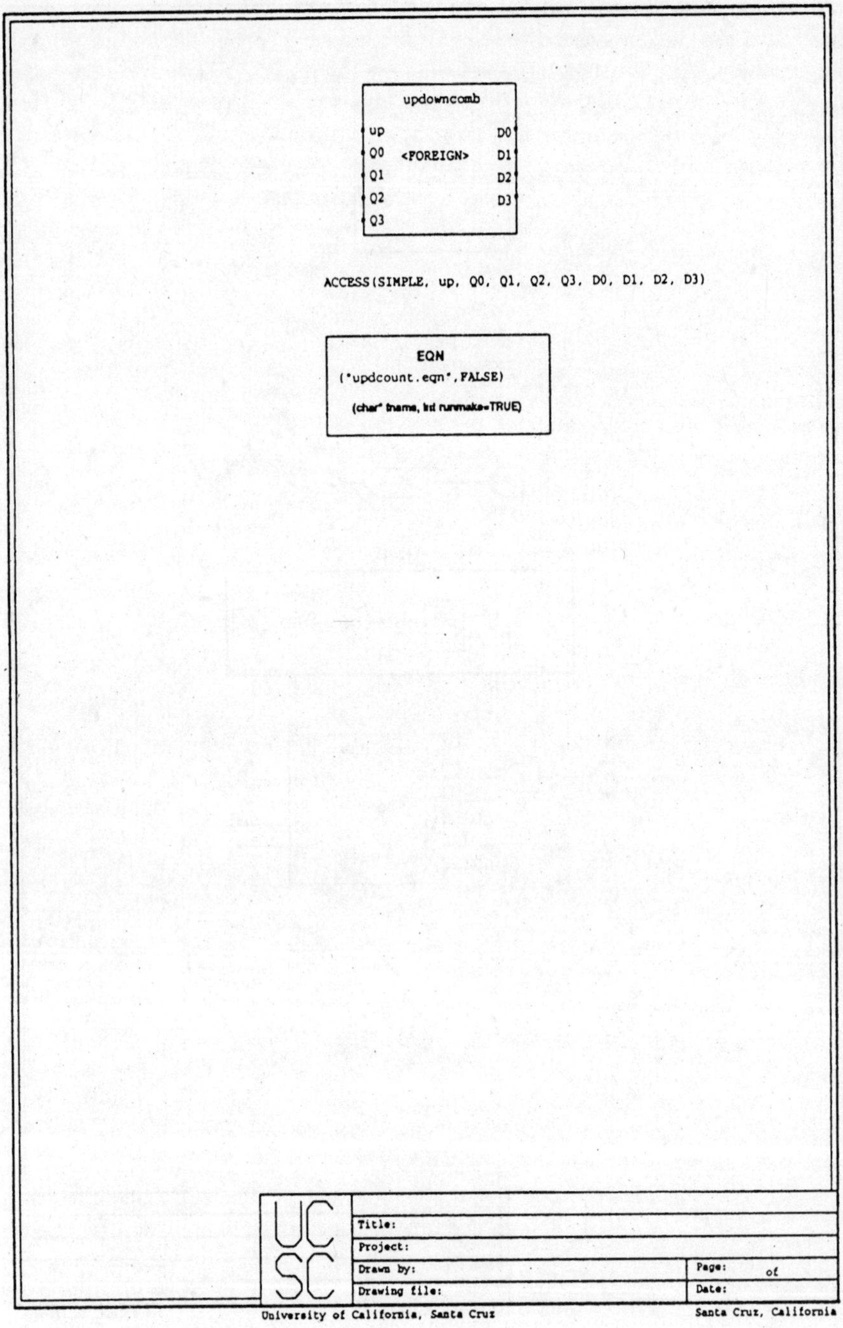

Figure 3.14: Up/Down Counter: Equation Symbol (Sheet 2 of 2).

The procedural capability of **wireC** allows other CAD tools to be incorporated into the design system. For example, there is a user-defined interface to the Berkeley logic synthesis system wherein the designer can represent part of the system behaviorally. **WireC** invokes logic synthesis on these components to produce a structural description that can be incorporated into the rest of the design. The simulator **xs** is available for a system-level simulation using a unit-delay model [11]. The **wireC** format is translated to **XNF** format. From here, the commercial proprietary tools take over: the **XNF** files are technology mapped by **xnfmap** and **map2lca** to **LCA** format and then are placed and routed by **apr**.

Example 3: We shall illustrate the design entry process taking this path with a simple up/down counter. The top-level diagram of the schematic drawing is depicted in Fig. 3.13.

This drawing specifies the clock input signal which goes through a global buffer **GCLK** to generate the buffered **clk** signal. The other input signal up uses an input pad and then an input buffer. The combinational logic is specified separately as **updowncomb**, while the flip-flops are included in a box and will be instantiated four times by the tool. The second and last sheet of this schematic drawing is shown in Fig. 3.14. The symbol in this sheet is mainly for signifying that the combinational logic **updowncomb** is coming from the equation file **updcount**.

3.4 Low-Level Design Entry

This involves using the XILINX **xact** design editor to enter a design at the "low" level, in which you have absolute control of the utilization of the resources available on an FPGA. The contents of the CLBs and the connection between the CLBs can be defined in the graphic design editor **xact**. An electronic image of a XILINX XC2064 FPGA as presented by **Xact** is given in Fig. 3.15.

With **xact**, the contents of the configurations of CLBs and IOBs have to be entered manually one by one, and the connections between the logic blocks are pretty much determined by the designer with only limited aid from a maze router. Manual design at such a low level has the merit that the utilization and performance of the design are high. There are designers who consider this path to be too time consuming. Unfortunately, there are times that this path is the only way to get the job done in the XC3000 series. The XC4000 series tools has a number of enhancements in the form of hard macros to hybridize manual and automatic implementation processes. The content and placement of a hard macro can be defined by a user in **xact**. It can then be tiled, duplicated automatically, and incorporated as components in the schematic drawing tool. Please consult the XILINX documentation *XC4000 Hard Macro Style Guide* for details.

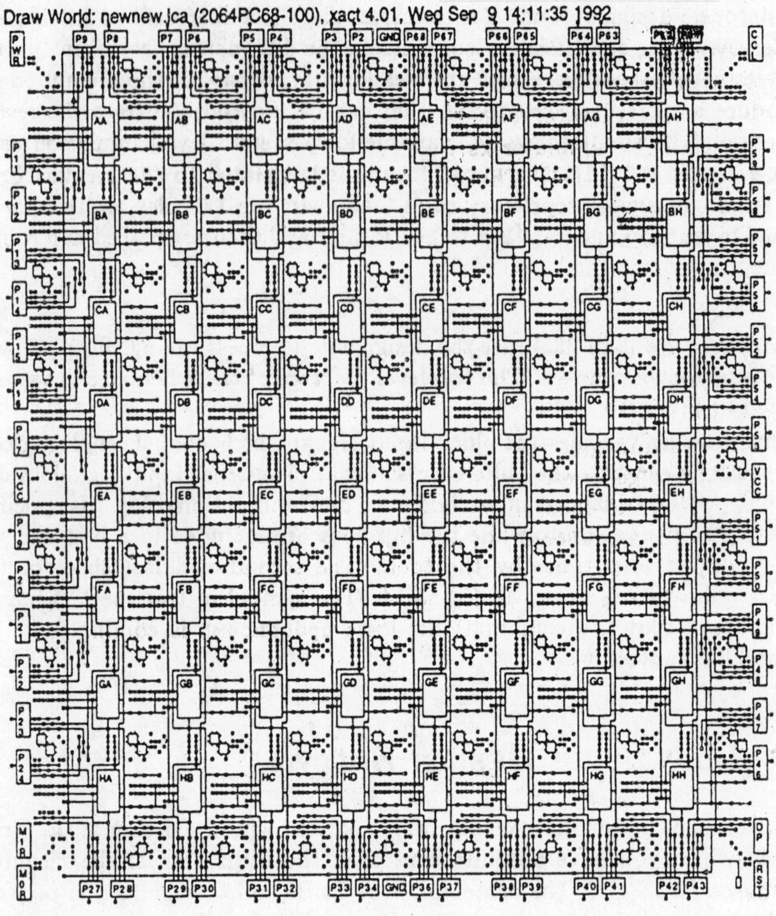

Figure 3.15: Xact Design Editor Model of a XILINX XC2064 FPGA.

3.5 Design Verification

Design verification tools are pretty generic in the sense that the verification process and user interface are generally not specific to any vendors, even though the specific performance/timing of the components are. For example, the verification process can start as early as right after design entry. If one assumes a unit-delay timing model for each component/gate, one can verify the functionality of the design at that level. On the other hand, to provide the timing verification tool with vendor specific data, the information regarding the delay of interconnects and components of an embedded design in a target technology, translator tools, such as XILINX `lca2xnf` and `bax`, are needed to extract timing information and *back annotate* the information for timing simulation.

A number of simulators are available for FPGA technologies such as `silos`, `viewsim`, and `susie`. The FPGA vendors themselves do not produce the simulation tools. The tools are produced by third-party vendors. Both `silos` and `viewsim` are "batch mode" simulators, meaning that you need to submit a sequence of commands (e.g., number of clock steps, assert this signal high or low) to the simulator ahead of time. But `susie` interacts with users interactively. `Susie` can simulate a design at the functional level with an `XNF` format file or at the timing level with an `LCA` format file after a design has been placed and routed. For details, see `susie` and XILINX development system manuals, Volumes I and II.

3.6 Design Realization and Prototyping

For one-time-programmable FPGAs like ACTEL, one needs a special device programmer to "blow" the antifuses inside the FPGA to configure the FPGA to realize the design.

For reprogrammable FPGAs like XILINX FPGAs, the `makebits` program are used to generate a bitstream to program/configure the FPGA. An FPGA device can be programmed in different modes. Serial mode is the most recommended way for the initial phase cf the prototyping. It is because the bitstream can be downloaded via a serial port of your computer with a download cable directly to the device. If your FPGA design is to be realized on board, you can use EPROMs, PROMs, or EEPROMs as a semipermanent way of providing the bitstream to configure the FPGA.

There are several methods available for realizing your FPGA-based design.

1. Breadboarding. This is the classical way of verifying a design using breadboards. Unfortunately, even though the lowend FPGAs do come in dual-in-line (DIP) packages (e.g., XILINX `XC2064PD48`), all the high-end FPGA parts are in packages that do not fit directly into a 0.1-inch-pitch standard breadboard. Special adaptors for the parts are required. For example, PLCC to DIP adaptors are available from JDR Microdevices [3] and Twin Industries, Inc. [4]

2. Vendor Demonstration Boards. These demo boards typically contain one FPGA, equipped with 7-segment displays, light emitting diodes, a small proto-area, dip switches, and momentary switches for demonstrations of simple designs.

3. Protozone Prototyping Board. These protoboards are originated from Stanford University [6]. A protozone kit comes in two pieces: a host card

[3] 2233 Samaritan Drive, San Jose, CA 95124.

[4] Twin Industries, Inc., 2921 Corvin Drive, Santa Clara, CA 95051.

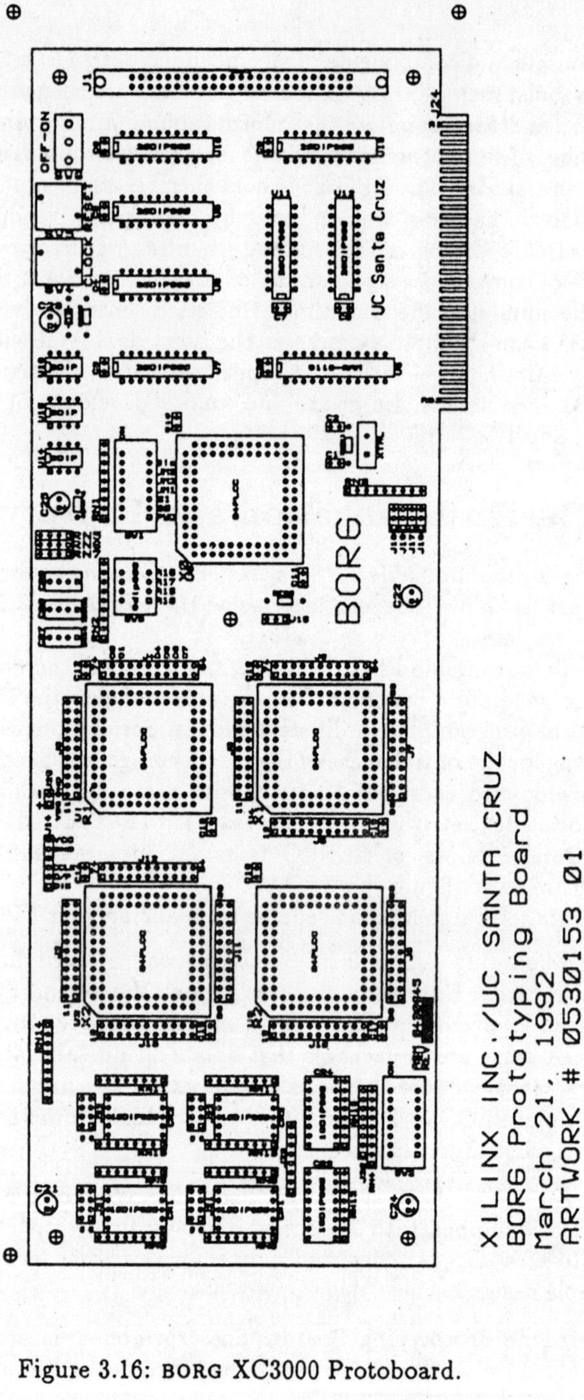

Figure 3.16: BORG XC3000 Protoboard.

and a protoboard. The host card is an "add-in" to a personal computer expansion slot. PC signals are derived/extended through a ribbon cable to reach the protoboard. The components on the protoboard have to be connected by wire-wraps. There is only one FPGA socket which comes with the protoboard. For details, see Protozone user's guide.

4. BORG Prototyping Board. This reusable protoboard, as illustrated in Fig. 3.16, is originated from University of California at Santa Cruz [3].[5] It is compatible with the Stanford protozone host card. The BORG is a PC-based prototyping board with two "user" FPGAs, two "routing" FPGAs, and a fifth FPGA that implements the glue logic to the PC bus. It exploits the reprogrammability of FPGAs and uses them for routing. It also comes with a toolset that facilitates the download and guides the process of using the FPGAs for routing.

Bibliography

[1] XILINX: *The Programmable Gate Array Data Book*. 2100 Logic Drive, San Jose, CA 95124, 1993.

[2] Brayton, R. K., R. Rudell, A. Sangiovanni-Vincentelli, and A. R. Wang. MIS: A Multiple-Level Logic Optimization System. *IEEE Transactions on Computer-Aided Design of Integrated Circuits and Systems*, CAD-6(6):1062–1081, Nov. 1987.

[3] Chan, P. K., M. Schlag, and M. Martin. BORG: A reconfigurable prototyping board using Field-Programmable Gate Arrays. In *Proceedings of the 1st International ACM/SIGDA Workshop on Field-Programmable Gate Arrays*, pages 47–51, Berkeley, California, USA, Feb. 1992.

[4] Devadas, S., H.-K. Ma, A. R. Newton, and A. Sangiovanni-Vincentelli. MUSTANG: State assignment of finite state machines targeting multilevel logic implementations. *IEEE Transactions on Computer-Aided Design of Integrated Circuits and Systems*, CAD-7(12):1290–1299, December 1988.

[5] Ebeling, C., and Z. Wu. WireLisp: Combining graphics and procedures in a circuit specification language. In *IEEE International Conference on Computer-Aided Design ICCAD-89*, pages 322–325, Santa Clara, CA, 5–9 November 1989. IEEE Computer Society Press.

[6] El Gamal, A. Protozone: The PC-Based ASIC Design Frame, User's Guide. Technical Report SISL90-???, Stanford Information Systems Laboratory, Stanford University, Aug. 1990.

[5] Please call or write to University Program Coordinator, XILINX Inc. 2100 Logic Drive, San Jose, CA 95124, for details.

[7] Fiduccia, C. M., and R. Mattheyses. A linear-time heuristic for improving network partitions. In *ACM IEEE* 19th *Design Automation Conference Proceedings*, pages 175–181, Las Vegas, Nevada, June 1982.

[8] Kernighan, B., and S. Lin, "An efficient heuristic procedure for partitioning graphs," *Bell System Technical Journal*, vol. 49, pp. 291-307, 1970.

[9] Kong, J., M. Schlag, and P. K. Chan. XNFWIREC tutorial: the Wirec manual derivative. Technical report, University of California, Santa Cruz, June 1991.

[10] Segal, R. B. *BDSYN Users' Manual Version 1.1.* University of California, Berkeley, University of California, Berkeley, 1989.

[11] Zien, J., J. Kong, P. K. Chan, and M. Schlag. XS - XILINX 2000/3000 FPGA Simulator. Technical report, University of California, Santa Cruz, Oct. 1991.

Chapter 4

Review of Logic Design

4.1 Overview

In this chapter we shall discuss some logic design principles and practices that are of relevance to designing with field-programmable gate arrays (FPGAs). As in the rest of this book, we assume that the reader has at least knowledge equivalent to a first course in logic design. The content of this chapter is considered as a review to prepare the reader for logic synthesis topics that will be covered in Chapters 5 and 6.

First, we shall review some basic concepts of logic functions and their implementations in two-level networks. In Section 4.3, the concept of *complete* functions and universal gates is explained. This will help in understanding the type of logic cell used by the different FPGAs. Issues related to minimization of combinational functions are discussed in Section 4.4. In Section 4.5, the Quine-McCluskey method is presented in a new form and as a prelude to logic synthesis to be presented in Chapter 5.

4.2 Boolean Algebra

A logic function may be defined in a truth table form or as a Boolean expression. The basic logical operations AND, OR, and NOT form a *complete* set in the sense that any logical function can be completely defined in these operations.

The function $f(a,b,c,d)$ is expressed as a sum of minterms:

$$f(a,b,c,d) = \sum m(3,4,6,7,12-15) \tag{4.1}$$

These minterms form the ON-set of the function. The minterms for which the function is zero form the OFF-set. In terms of the variable, this function can be written as

$$f(a,b,c,d) = a'b'cd + a'bc'd' + a'bcd' + a'bcd + abc'd' + abc'd + abcd' + abcd \tag{4.2}$$

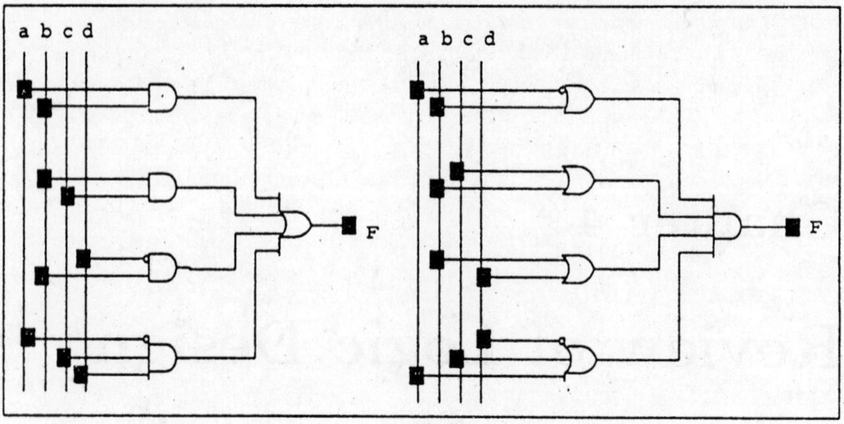

Figure 4.1: Two-Level Logic Representation.

The function may be simplified and expressed as *sum of products* (SOP):

$$f(a,b,c,d) = ab + bc + bd' + a'cd \tag{4.3}$$

There are several techniques to minimize a function. Algebraic minimization makes use of the Boolean algebra axioms and fundamental theorems. Other techniques such as Karnaugh maps and the Quine-McCluskey tabular method are discussed in a later section of this chapter and in Chapter 5.

The function $f(a,b,c,d)$ can also be expressed as a *product of sums* (POS). This can be obtained by repeatedly applying to the SOP expression of Eq. (4.3) the distributive law on OR, $x + yz = (x+y)(x+z)$, and simplifying the resulting expressions. We obtain

$$f(a,b,c,d) = (b+a')(c+b)(a+c+d')(b+d) \tag{4.4}$$

This expression can also be obtained starting from the POS form of the function

$$f(a,b,c,d) = \prod M(0,1,2,5,8-11) \tag{4.5}$$

Figure 4.1 shows the SOP and POS implementations of the function $f(a,b,c,d)$.

Exercise 1: Find the POS of $f(a,b,c,d)$ by taking the dual of Eq. (4.3), multiplying, then taking the dual again. Verify with the logic expression of Eq. (4.4).

4.3 Designing with Complete Gates

Usually logical functions have been expressed in terms of the complete set of operations AND, OR, and NOT. The gates implementing these operations are also called a *complete set of gates* [2]. However, these gates, which are extensively used in logic design and schematic capture do not have equivalent actual circuits in most technology. For example, in n-channel metal-oxide semiconductors (NMOS) and complementary metal-oxide semiconductors (CMOS) technologies, the NOR and NAND gates are the preferred gates. Fortunately, there are other complete sets of gates that can be used to implement logical functions. Among these gates are NAND, NOR, XOR, and AND; multiplexer (or tree of multiplexers); and random-access memory (RAM) with decoding circuitry. XOR and AND are Modulo 2 algebraic operations [3].

To prove that a set of gates is complete, we need to show that this set can implement the NOT, AND, and OR operations. This is illustrated in Fig. 4.2 for the most popular complete sets that are used as logic modules in FPGAs. For example, Actel and QuickLogic use multiplexers; Xilinx uses RAM cells; Concurrent uses XOR and AND gates; and Plessy uses NAND gates. The smaller the module, the finer is the granularity of the FPGAs. Plessy has finer granularity (NANDs) than does Actel (4:1 multiplexers), and Xilinx's 32-bit RAMs are the coarsest grain of the three.

Exercise 2: Show that the set of gates OR and XOR is complete.

4.3.1 Implementation with NAND and NOR Gates

A Boolean function can be expressed exclusively in terms of NAND or NOR operations. Using DeMorgan's theorem, $(xy)' \equiv x' + y'$, the expression in Eq. (4.3) can be expressed in NAND/NAND gates:

$$f(a,b,c,d) = [(ab)'(bc)'(bd')'(a'cd)']' \qquad (4.6)$$

Similarly, the same function can be expressed in NOR/NOR starting with the expression in Eq. (4.4) and applying DeMorgan's theorem to obtain

$$f(a,b,c,d) = [(b+a')' + (c+b)' + (a+c+d')' + (b+d)']' \qquad (4.7)$$

4.3.2 Designing with Multiplexers

For the material presented in this section and in the next chapter, we need to quickly review the concept of *residues*.

Definition: The residue of a function $f(x_1, x_2, \ldots, x_n)$ with respect to (wrt) a variable x_j is the value of the function for a specific value of x_j. It is denoted by f_{x_j} for $x_j = 1$ and by $f_{x'_j}$ for $x_j = 0$.

The residues, wrt a, of $f(a,b,c,d)$ defined in Eq. (4.3) are $f_{a'} = bc + bd' + cd$ and $f_a = b$. The function can then be expressed in terms of these residues:

$$f = a' f_{a'} + a f_a$$

This method of decomposing a function is also known as Shannon's decomposition. The concept of residues can be extended to more than one variable.

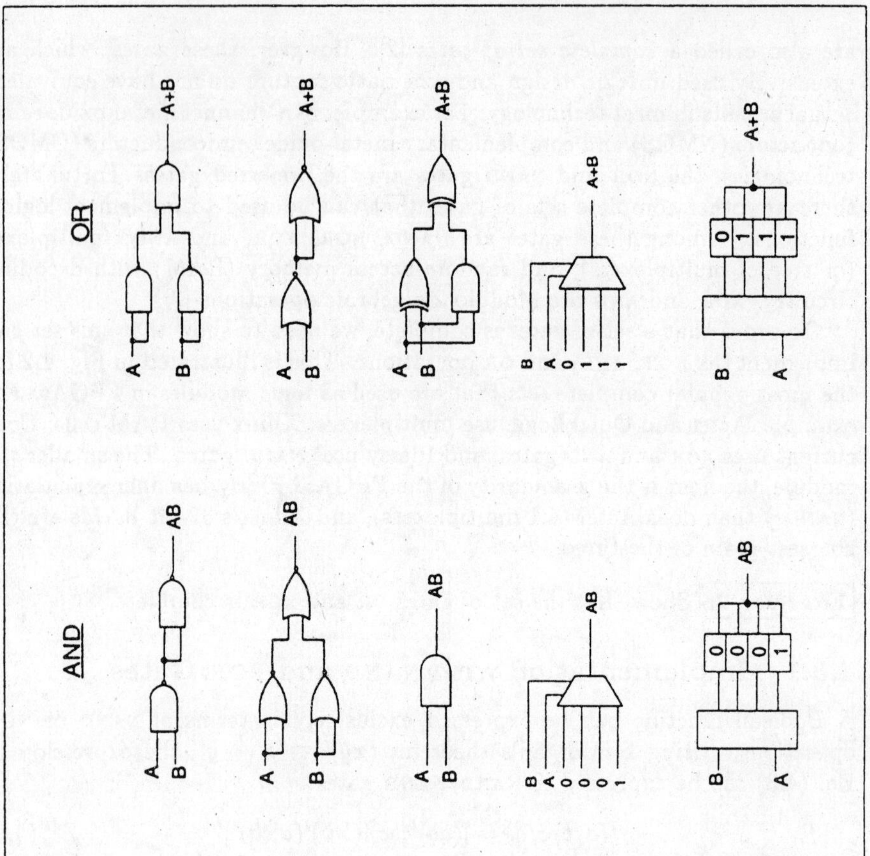

Figure 4.2: Complete Sets.

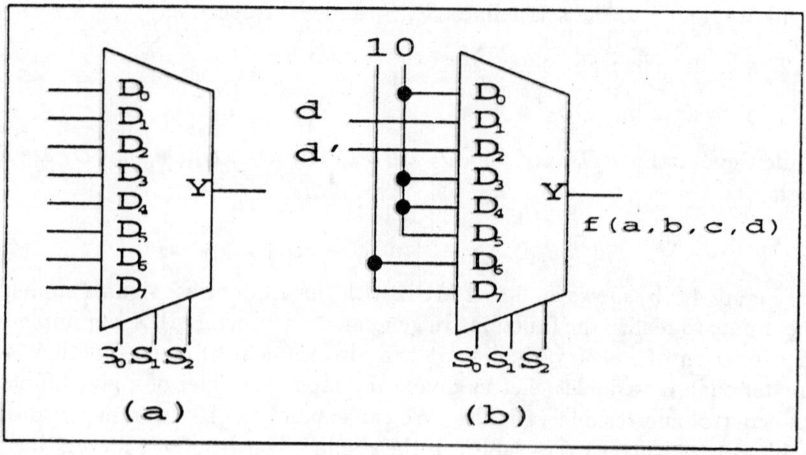

Figure 4.3: Multiplexers as Function Generators.

Thus $f_{ab} = 1$ is the residue of $f(a,b,c,d)$ for $a = b = 1$. In general we shall denote the residues by R with numeric subscripts that represent the values of the variables. Thus R_3 is equivalent to f_{ab} and R_0 is equivalent to $f_{a'b'}$, and so on.

Consider the 8-to-1 MUX in Fig. 4.3. The output of this MUX can be expressed in terms of the input data $D_0, \ldots, D_7$ and the three selection lines $S_0S_1S_2$ as

$$f = m_0D_0 + m_1D_1 + m_2D_2 + m_3D_3 + m_4D_4 + m_5D_5 + m_6D_6 + m_7D_7 \quad (4.8)$$

$$f(a,b,c,d) = \sum m_iD_i \quad (4.9)$$

where m_i is the ith minterm of S_0, S_1, and S_2.

On the other hand, the function $f(a,b,c,d)$ given in Eq.(4.1) can be rewritten in terms of its residues wrt a,b,c in the form

$$f(a,b,c,d) = m_0R_0 + m_1R_1 + m_2R_2 + m_3R_3 + m_4R_4 + m_5R_5 + m_6R_6 + m_7R_7 \quad (4.10)$$

$$f(a,b,c,d) = \sum m_iR_i \quad (4.11)$$

where R_i is the residue of the function for the values of a, b, and c corresponding to the ith minterm. The comparison of Eqs. (4.8) and (4.10) shows that the function $f(a,b,c,d)$ can be implemented in the 8-to-1 MUX provided the data lines D_i assume the same values as the corresponding residue R_i of the function. Here, $R_0 = R_4 = R_5 = 0$, $R_1 = d$, $R_2 = d'$, and $R_3 = R_6 = R_7 = 1$. Thus we apply the inputs

1. a, b, and c to the select lines S_0, S_1 and S_2, respectively
2. d to data line D_1
3. d' to data line D_2

while the data lines D_0, D_4, and D_5 are held low and D_3, D_6, and D_7 are held high.

$$f(a,b,c,d) = m_1 d + m_2 d' + m_3 + m_6 + m_7 \tag{4.12}$$

Figure 4.3(b) shows an 8-to-1 MUX with the appropriate signals applied on the inputs to realize the function. In general, an n-control MUX can implement any function of $n+1$ variables. It can also implement some functions with greater than n variables. Let us divide the input variables of a given function into control and residue variables. We can expand the function in terms of its residues wrt the control variables. If the residues consist of one literal, then the function can be implemented by the n-control MUX. For example, the 8-to-1 MUX, which has three control lines, may be used to implement the function of five variables:

$$g(a,b,c,d,e) = ab + cd + a'b'c'e + abc'e \tag{4.13}$$

which can be written in terms of a, b, and c and its residues for the same variables as:

$$g(a,b,c,d,e) = abc + abc' + a'b'c'e + a'b'cd + a'bcd + ab'cd + abcd + abc'e \tag{4.14}$$

$$g(a,b,c,d,e) = (a'b'c')e + (a'b'c)d + (a'bc)d + (ab'c)d + (abc') + (abc) \tag{4.15}$$

Applying the residues of the function to the data lines of the MUX yields the implementation shown in Fig. 4.4.

However, the function

$$h(a,b,c,d,e) = ab + cd + e \tag{4.16}$$

cannot be implemented using only one of these MUXs. In Fig. 4.4(b) some of the data lines are functions of d and e. Extra gates are needed to implement these functions. Similarly, if we attempt to realize the function f given by Eq. (4.2) using one 4-to-1 MUX by applying the variables a and b to the control lines, we immediately find that it is not possible simply to apply individual variables to the data lines.

$$f(a,b,c,d) = a'b'cd + a'b(c + d') + ab \tag{4.17}$$

The Karnaugh map of $f(a,b,c,d)$ is shown in Fig. 4.5. Each horizontal line represents a minterm of a and b. The corresponding residue is indicated. Now the inputs on data line D_0 and D_1 are functions that can themselves be implemented using MUXs as shown in Fig. 4.5. Such an iterative process allows us to decompose the function of several variables into functions of fewer variables.

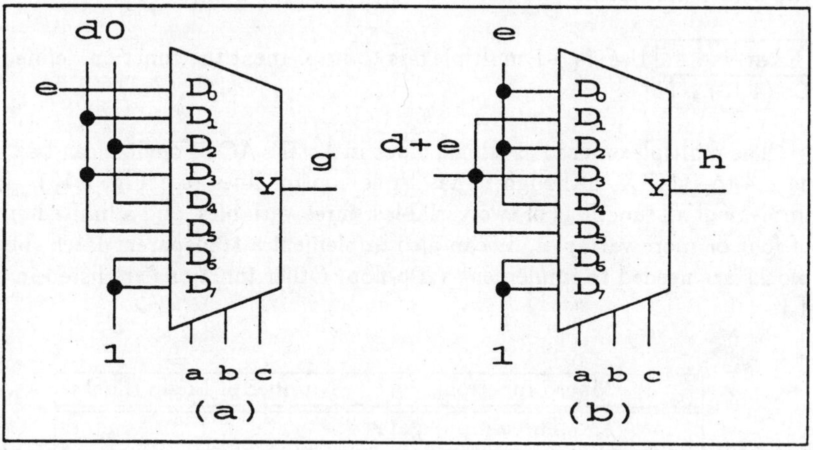

Figure 4.4: Implementations of Five-Variable Functions with an 8-to-1 MUX. (a) Function g Given by Eq. (4.15); (b) Function h Given by Eq. (4.16).

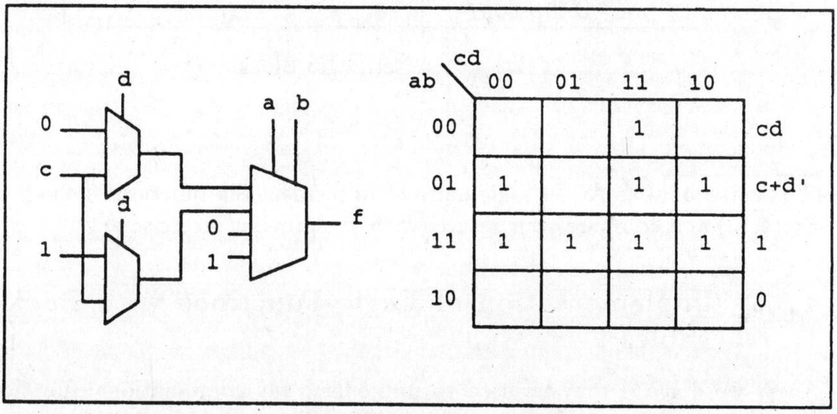

Figure 4.5: Decomposition of the Function Shown in Fig. 4.4.

Exercise 3: Use a 2-to-1 multiplexer to implement a D-latch. Then, use two 2-to-1 multiplexers to build an edge-triggered D-flip-flop.

Exercise 4: Use 4-to-1 multiplexers to implement the function defined by Eq.(4.13).

The multiplexer used as a logic block in Actel's ACT1 devices can be viewed as a 4-to-1 MUX, although it has three control lines (see Fig. 2.13). It can implement all functions of two variables, three variables, and a limited number of four or more variables. It can also implement a transparent latch, but two blocks are needed to implement a flip-flop. Other functions are listed in Table 4.1.

Macro function	Number of Logic Blocks
Two- and Three-input gate	1
Four-input gate	2
XOR or XOR OR (AND)	1
AND/OR or OR/AND	1
Buffer/inverter	1
Two- or 4-input mux	1
Adder	2
D-latch, with clear or enable	1
Multiplexed D-latch	1
D-flip-flop	2

Table 4.1: Actel Logic Blocks.

Exercise 5: Use ACT1 logic units to implement the functions defined by Eq.(4.13) and compare your results with the previous exercise.

4.3.3 Implementation of Logic Functions with Look-up Tables

An N-word RAM may be used to implement any combinational function of $\lfloor \log_2 N \rfloor$ variables as a look-up table (LUT). The variables of the function are applied to the address lines of the RAM. The values of the function for all combinations of the inputs are stored in the corresponding words of the RAM. The concept is illustrated for the implementation of a two-input XOR function in Fig. 4.6.

There is quite a similarity between such a representation and implementation in a MUX. In Fig. 4.6, the XOR function is used to illustrate the analogy. In the case of the MUX, the inputs a and b are applied to the selection lines

S_0 and S_1. The residues of the function with respect of a and b are applied to the data lines of the MUX: $R_0 = R_3 = 0$ and $R_1 = R_2 = 1$.

The RAM consists of four 1-bit words. The variables a and b of the XOR function are the address lines of the RAM. The values of $f = a \oplus b$ for each combination of a and b are stored in the corresponding bits of the RAM.

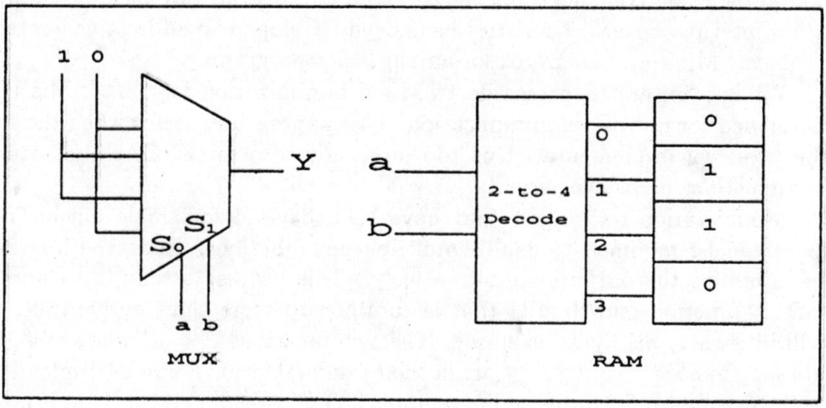

Figure 4.6: Analogy Between RAMs and Multiplexers.

4.4 Minimization of Combinational Functions

Functions expressed in truth tables or minterms need to be minimized prior to implementation. Minimization is performed to reduce

- Circuit size, gates and literals (space and power)
- Number of levels of logic (time or circuit depth)

Often the two objectives cannot be met simultaneously and some tradeoffs are necessary between *time* and *space*. For example, a two-level implementation of a function has shorter delays than a multilevel implementation. However, the former implementation may require a larger fanin (larger number of literals) than the latter does. Larger fanin gates may have larger gate delays. This is actually technology dependent. In CMOS, for example, the delay caused by high fanin is more significant than that resulting from multi-level logic. Multilevel minimization is particularly important to FPGAs which are multilevel logic circuits.

Regardless of the reasons for two-level minimization, there are several techniques used. The most common techniques are: Karnaugh maps [5], Quine-McCluskey (QM) method [8] [6], iterative consensus, and computer programs such as Mini [4], Presto [1], and Espresso.

Karnaugh maps are used for designs with a small number of inputs. The QM method is more amenable to programming. However, since it is based on exhaustive comparisons and merging, it demands an unreasonable amount of computer resources and time. A 20-variable function may have as many as 2^{19} minterms. Several heuristics have been developed to address the covering problem: Mini [4], and Espresso, among other programs.

While Mini and Espresso do two-level minimization for PLAs, MisII was developed for multilevel minimization. This approach is useful when the fanin for two-level implementation is too large and also for FPGAs that are, by construction, multilevel circuits.

Minimization techniques that have been devised for single output functions can be modified to handle multi-output functions. Here, the interest is to minimize the outputs simultaneously. This process attempts to find the largest function (subcircuit) that is common to more than one output. We will illustrate this by an example. The two functions $F = abcd + abc'd', G = abe+abf'+abcde'f+abc'd'e'f$ are in their minimal form. F can be implemented with 10 fanins and 3 gates and G requires 22 fanins and 5 gates. That is a total of 32 fanins and 8 gates. Multiple output minimization would allow us to find terms common to the two functions, mainly the two products of F. Thus G can be written in the form: $G = abe + abf' + (abcd + abc'd')e'f = abe + abf + Fe'f$ and the two functions can be implemented with 22 fanins and 7 gates which is a considerable reduction in literals and gates.

The advantages are even greater when this approach is used with FPGAs. For example, to realize these two functions in XC3000 devices, 3 CLBs are needed in the first and only 2 are sufficient in the second case. F is a function of four variables which can be implemented in one CLB, while G depends on 6 variables and needs two CLBs. When we expressed G in terms of F and only four of the variables, a, b, e, and f, the function becomes dependent on five variables and is then implemented in one CLB. Multilevel minimization does not necessarily yield results that are easily implemented to a specific technology. Special attention needs to be paid to tailor the minimization for a specific technology: MUX, RAM, NAND, etc. Grouping the minimized functions to fit into any of these modules is known as *technology mapping* and will be discussed in Chapter 5.

In the next section, we shall review minimization using QM's method in the form of an algorithm, and illustrate the use of this algorithm by examples. The rationale for taking this approach is that, more and more, synthesis of logic systems is automated. Actually, the word *synthesis* is today almost always used to refer to *automated digital design*. Developing an algorithm for a procedure with which the readers are familiar will help them follow the theory behind MisII described in Chapter 5.

4.5 The Quine-McCluskey Algorithm

The Quine-McCluskey (QM) method is a systematic procedure for the minimization of Boolean functions that are expressed as a sum of minterms [8]. It avoids relying on the visual perception used in the Karnaugh map method in forming *prime implicants*. The method consists of two main steps:

1. Determine the prime implicants of the function.
2. Find a minimal cover for the function.

First we define some sets and operators that we shall use to present the algorithm. We express an N-variable function, f, as a sum of minterms (0-cubes) in the form,

$$f = \sum_{j=0} C_j^0$$

where C_j^0 is a 0-cube. These cubes and higher-order cubes will be denoted by C_j^k, where j is an order index and $0 \leq k \leq N$ denotes the number of x's in the cube. Thus the 1-cubes are C_j^1 and the 2-cubes are C_j^2. Each k-cube can be expressed in binary form as $C_j^k = b_{N-1} \ldots b_j \ldots b_1 b_0$, where any b_i is 0, 1, or x, x being either 0 or 1. For example, if we join two 0-cubes of a function of three variables, (010) and (011) we obtain the 1-cube (01x). This operation is possible because the two 0-cubes differ in 1-bit position. They have a *distance* d of 1:

$$d(C_i^0, C_j^0) = 1 \tag{4.18}$$

The operation is based on the consensus principle, $ab + ab' = a$. We denote the operation of joining two n-cubes, of distance 1, to form an $(n+1)$-cube as the *union* of these n-cubes:

$$C_k^{n+1} = C_j^n \cup C_i^n \tag{4.19}$$

Any i-cube is also called an implicant.

Example 1: If C_j^2 = x1x1 and C_k^2 =x0x1, then the 3-cube is $C_l^3 = C_j^2 \cup C_k^2$ = xxx1.

4.5.1 Definitions

1. The *intersection* of two cubes, C_1 and C_2, is defined as the set of cubes common to both C_1 and C_2. The intersection can be obtained using the bit operations defined in Table 4.2. The operation is applied on the corresponding bits of the two cubes. If the result of the operation contains Ω, then the the two cubes do not intersect; that is, $C_1 \cap C_2 = \emptyset$.

∩	0	1	x
0	0	Ω	0
1	Ω	1	1
x	0	1	x

Table 4.2: The Intersection Operation.

	00	01	11	10
0			1	1
1			1	1

Figure 4.7: Karnaugh Map.

2. The *sharp operation* on two sets, A and B, is denoted by $A\#B$ and results in a set W that contains all elements of A that are not in B. Thus this operation can be written as $W = A \# B = A - (A \cap B)$.

Using this terminology and operations, we shall develop procedures to determine the prime implicants (PI), the essential prime implicants (EPI), and redundant terms. Then we shall show how these procedures will be used to find a minimal cover.

4.5.2 Determining the Prime Implicants

Let us recall that a PI is an implicant that cannot be a distance 1 apart from any other implicant of the function. That is, a PI cannot be combined with another implicant to eliminate a variable. Identifying the PI is based on the same principle used in the Karnaugh map method, using: $ab + ab' = a$. In the Karnaugh map method it is possible to join four 0-cubes into a 2-cube by visual inspection. For example, the 0-cubes, (010), (011), (110), and (111), can be visually combined in the 2-cube (x1x) as illustrated in Fig. 4.7. However, in the tabular method, the formation of PI is done in a more systematic and exhaustive manner [6]. Starting with a set of 0-cubes, all possible 1-cubes are formed, then all 2-cubes, and so on. As cubes are joined to form higher-order cubes, they are discarded. Continuing with the same example, first all 1-cubes, (01x), (11x), (x10), and (x11), are formed. The next step involves the formation of all 2-cubes. The 1-cubes (01x) and (11x) are joined to form (x1x), and the cubes (x10) and (x11) form (x1x) also. Clearly, the two 2-cubes are identical. In the QM method, as all PI are formed, it is important to make sure that there is no duplication of cubes. Thus we need to eliminate redundant cubes.

Let f be an N-variable function that is expressed as a sum of minterms:

$$f = \sum_{j=0} C_j^0 \tag{4.20}$$

The first step in the QM method is to group the 0-cubes C_j^0 according to the number of 1's in their binary representation. Thus the 0-cube (0000) is in a group by itself, no 1's. Cubes such as (0001), (0010), (0100), and (1000) form the next group; those that contain two 1's form the subsequent group; and so on. Let S^k be the set of all k-cubes and $S_i^k \subset S^k$ be its subsets that contain the k-cubes that have the same number of 1's. The index i indicates the number of 1's. Then the sets S_i^k form a partition of S^k. We denote the cardinalities of these sets by $\| S_i^k \|$ and the number of these sets by n_k. To form higher-order cubes, only cubes of two "adjacent" groups need to be examined since the distance of two cubes from nonadjacent groups will be larger than 1. To find the prime implicants, the distance of k-cubes from adjacent subsets will be examined, and if it is equal to 1, the k-cubes will be joined to form a new $(k+1)$-cube, C^{k+1}.

1. Form the sets S_i^0
2. Form the PI:

```
For k = 0 to N − 1
    For i = 0 to N − k − 1
        For each C_j^k ∈ S_i^k
        For each C_m^k ∈ S_{i+1}^k
                If d(C_j^k, C_m^k) is 1 then begin
                        Create new cube C^{k+1} = C_j^k ∪ C_m^k,
                        Put C^{k+1} ∈ S_i^{k+1},
                        Tag C_j^k and C_m^k,
                        end
        next C_m^k
        next C_j^k
        If S_i^{k+1} ≠ ∅ add S_i^{k+1} to S^{k+1}
    next i
    If S^{k+1} = ∅ then done
next k
```

Next we discard all tagged cubes, and the remaining cubes form the set of prime implicants P. We illustrate the use of this algorithm with an example. The traditional way of determining PIs using tables is depicted in Table 4.3.

Example 2:

Find the prime implicants for the function $f(a, b, c, d) = \sum m(0, 1, 3, 6, 7, 14, 15)$

0-cube	1-cube	2-cube
0 ✓	0,1	6,7,14,15
1 ✓	1,3	
3 ✓	3,7	
6 ✓	6,7 ✓	
7 ✓	6,14 ✓	
14 ✓	7,15 ✓	
15 ✓	14,15✓	

Table 4.3: Traditional Tabular Method.

- Construct the subsets S_i^0:
  $S_0^0 = (0) = (0000)$
  $S_1^0 = (1) = (0001)$
  $S_2^0 = (3, 6) = (0011, 0110)$
  $S_3^0 = (7, 14) = (0111, 1110)$
  $S_4^0 = (15) = (1111)$
  then n_0=5, $\mid S_0^0 \mid = \mid S_1^0 \mid = \mid S_4^0 \mid = 1$, and $\mid S_2^0 \mid = \mid S_3^0 \mid = 2$.
- Next form S^1 sets:
  $S_0^1 = (0, 1) = (000x)$
  $S_1^1 = (1, 3) = (00x1)$
  $S_2^1 = (3, 7)\ (6, 7)\ (6, 14) = (0x11, 011x, x110)$
  $S_3^1 = [(7, 15), (14, 15)] = (x111, 111x)$
  Tag all C_i^0, (all 0-cubes joined in 1-cubes).
  then n_1=4, $\mid S_0^1 \mid = \mid S_1^1 \mid = 1$, and $\mid S_2^1 \mid = \mid S_3^1 \mid = 2$.
- Next, find the S^2 subsets:
  $S_0^2 = [(011x, 111x), (x110, x111)] = [(x11x), (x11x)]$
  Notice that the 2-cube (x11x) is entered twice in set S^2. One of these two entries will be discarded. Also, we discard all tagged cubes which are 011x, x110, x111, and 111x. The remaining five cubes (000x, 00x1, 0x11, x11x, x11x) form the prime implicants of the function.

Exercise 6: Use the algorithm described in this section to find the PI of $f(a, b, c, d)$ as defined in Eq. (4.1).

4.5.3 The Minimal Cover

A cover, C, of a function consists of prime implicants (essential and, if necessary, secondary) that contain all the 0-cubes of the function. A cover is minimal if it does not contain another cover. A paper and pencil method places all the PIs in a table known as the PI table. From this table, an EPI is identified and then removed from the table as well as all the cubes that form this EPI. The reduced table is then simplified by *row covering*, a process that can be recognized as eliminating redundant terms. Before proceeding with the compilation of a minimal cover, we shall define some procedures.

Let P be the set of prime implicants of a function whose minterms form the set C. To determine the minimal cover, we need to define three procedures on sets: *Identify EPI*, *Update-Set*, and *Remove-Redundant.* This last procedure is equivalent to row covering. It is used to remove dominated rows from the PI table.

1. *Identify EPIs.* An EPI is a PI that contains cubes that do not belong to another PI. We absolutely need the EPI to cover the minterm(s). If there is a PI p_i that is covering a minterm that is not covered by any other PI, then p_i is essential. We identify p_i by using the sharp operator. Thus we shall start with two PIs, $p_i \in P$ and $p_j \in P$, and find the 0-cubes of p_i that do not belong to p_j : $(\alpha = p_i \# p_j)$. Then we check if any of them belongs to any other PI in the set. If not, p_i is an EPI. The procedure can be expressed in the following form:

   Procedure *Identify-EPI(p_i)*

   $\alpha = p_i$

   For each $p_j \neq p_i; \quad p_j \in P$

   $\alpha = \alpha \# p_j$

   if $\alpha = \emptyset$ **then done** p_i **is not** EPI

   end

   p_i is an EPI

2. *Update-Sets.* As soon as an EPI is identified, it is removed from the P set. Also all 0-cubes in this PI are removed from the C set and from all other PIs.

   Procedure *Update-Set (P, C)*

   $P = P \# p_i$, $C = C \# p_i$, and $p_j = p_j \# p_i$, **for all** $j \neq i$

3. *Remove-Redundant.* An implicant is redundant if it is contained in any other implicant. That is, p_i is redundant if there is at least one member $p_j \in P$ that totally contains p_i. In terms of the intersection, a prime implicant p_i is redundant if $p_i \cap p_j = p_i$. Recognition of redundant terms is important in simplifying the PI table. As an EPI is identified, it is deleted from the table and so are all 0-cubes that form this EPI. At this stage, the QM method uses row covering [7] to eliminate some of the remaining prime implicants.

   Procedure *Remove-Redundant*

   For $p_i \in P$ **and** $p_j \in P$

   If $p_i \cap p_j = p_i$ **for any** $j \neq i$ **then** p_i **is redundant**

4. *Process Cover(P).*

   - Step 1: *Remove-Redundant*

- **Step 2: Last = $\| P \|$**

```
For p_i ∈ P,
    Identify-EPI(p_i)
    If p_i is an EPI then begin
            Update-Set(P,C)
            Remove-Redundant
            If C = ∅ then done.
                Else If (C ≠ ∅ and | P |= 1) then done
                    Else If i = Last
                    then P is cyclic
            end
next p_i
```

Example 3: Continuing with the example in the previous section for which we found that the function has five PIs, we noticed at the end of that process that two of these PIs were identical. We show in Table 4.4 the traditional prime implicant table that is used to find the cover.

- $P = (p_1, p_2, p_3, p_4, p_5)$
- *Remove-Redundant*
  $p_4 \cap p_5 = p_4 = p_5$, so remove p_5 (or p_4).

- *Identify-EPI*
  $p_1 = (000x)$, $p_2 = (00x1)$; 'x' can be 0 or 1
  $p_1 \# p_2 = (0000)$
  $(p_1 \# p_2) \cap p_k = \emptyset$ for all $k = 3$ to 4
  p_1 is an EPI.

- *Update-Set* C and P
  $C = C \# p_1 = (0011, 0110, 0111, 1110, 1111)$, and
  $P = (p_2, p_3, p_4)$ where
  $p_2 = p_2 \# p_1 = (0011)$
  $p_3 = p_3 \# p_1 = (0x11)$ (remains unchanged)
  $p_4 = p_4 \# p_1 = (x11x)$ (remains unchanged)

- *Remove-Redundant* implicant
  $p_2 \cap p_3 = p_2$, Then p_2 is redundant
- *Identify-EPI*:
  $p_3 \# p_4 = (0011)$ (not empty);
  Then p_3 is an EPI

- *Update Sets*
  $C = C \# p_3 = (0110, 1110, 1111)$ and
  $P = P \# p_3 = (x11x) = p_4$.
  Since $| P | = 1$, then p_4 is an EPI and all EPIs are found.

PI	0	1	3	6	7	14	15
0, 1	x	x					
1, 3		x	x				
3, 7			x		x		
6, 7, 14, 15				x	x	x	x

Table 4.4: Prime Implicant Table.

	m_0	m_1	m_2
p_1	x	x	
p_2	x		x
p_3		x	x

Table 4.5: An Example of Cyclic PI Table.

To express the minimal cover of the function in terms of the variable, $a, b, c, \ldots$, we replace every 0 and 1 in the implicants of the cover by the corresponding variable, complemented or noncomplemented, respectively.

4.5.4 Cyclic Covers

In the case of cyclic covers, as illustrated in Table 4.5 we break the cycle by eliminating arbitrarily any prime implicant, p_i. This is justified because every 0-cube in p_i belongs also to other PIs (otherwise, p_i is an EPI). Once the cycle is broken, we can proceed with the cover algorithm. It is possible that another cyclic configuration is again encountered, and it can be broken down in a similar fashion. Another approach to breaking the cycle was recommended by McCluskey [7].

4.5.5 Functions with Don't Care

If the function to be minimized contains Don't Care terms, then these terms are treated as are the other minterms when PI are formed. However, they are not included in the minterms when the cover is determined.

Exercise 7: Apply the QM algorithm for the incompletely specified function (including Don't Care):

$$f(a, b, c, d) = \sum m(3, 4, 6, 7, 12 - 15) + d(1, 5) \tag{4.21}$$

4.6 Summary

This chapter presented a review of combinational logic design with emphasis on features relevant to designing with FPGAs: complete sets, implementation with MUXs and RAMs. The chapter pointed out the need for an algorithmic approach to solve minimization of combinational functions in two- and multilevel. Heuristics are used to solve these covering problems. Algorithms that are based on these heuristics are used in automated design tools. The algorithmic approach to solve these complex problems is illustrated with the Quine-McCluskey method which is familiar to most readers.

Bibliography

[1] Brown, D. W. "A state machine synthesizer-SMS," *Proc. of the 18th Design Automation Conference*, pp. 301-305, June 1981. Prentice-Hall, 1991.

[2] Chu, Y. *Digital Computer Design Fundamentals.* New York: McGraw-Hill, 1962.

[3] Green, D. *Modern Logic Design.* Reading, MA: Addison-Wesley, 1992.

[4] Hong, S. J. , R.G. Cain, and D. L. Ostapko, "MINI: A heuristic approach for logic minimization," *IBM J. of R&D*, Vol. 18, No. 5, pp. 443-458, Sep. 1974.

[5] Karnaugh, M. "The map method for synthesis of combinational logic circuits," *AIEE Trans. in Comm. and Electr.*, 72, pp. 593-599, Nov. 1953.

[6] McCluskey, E. J. "Minimization of Boolean functions," *Bell System Technical Journal*, Vol. 35, no. 5, pp. 1417–1444, 1956.

[7] McCluskey, E. J. *Logic Design Principles with Emphasis on Testable Semicustom Design.* Englewood, Cliffs, NJ: Prentice Hall, 1986.

[8] Quine, W. V. "The problem of simplifying truth functions," *American Mathematics Monthly*, Vol. 59, pp. 521–531, 1952.

Chapter 5

Multilevel Logic Minimization

5.1 Introduction

Logic minimization is an important phase in the design flow that attempts to simplify the design to its "minimal" form for realization. In this chapter, we focus on combinational logic minimization, as opposed to sequential logic minimization.

Currently, Xilinx tools lack logic minimization capability, and commercial tools from Synopsys and others are expensive. The first author incorporates university logic minimization tools (`misII` and `espresso`) [4] into a classroom environment with some success.

A word of caution: logic minimization tools must be used by those who understand them, so as not to abuse them. Quite frequently, users discover their results are useless after spending hours and hours of computer time with the logic minimizer. To reduce futile efforts, one should know some of the algorithms behind them and their limitations to use these tools meaningfully and effectively. The purpose of this chapter is to provide some theory and hands-on experience with a particular Boolean minimization tool: `misII`. Later in this chapter, we shall also examine the relationship between Boolean minimization and technology mapping.

At UC Santa Cruz, the interface of `misII` to Xilinx tools is provided by several utilities devised by Jackson Kong: `eqn2xnf` and `mapff`.[1] The first utility translates Boolean equation format to Xilinx Netlist Format, which is the input format of Xilinx technology mappers. `Mapff` appends flips-flops to the combinational part of a finite state machine to complete the design without drawing schematics.

[1] Available upon request. Please write to the first author at: 225 Applied Sciences, University of California at Santa Cruz, CA 95064.

5.2 Representation of Boolean Functions

5.2.1 Truth Tables, ON-Set, OFF-Set, DC-Set

Truth tables are the most general, unsophisticated representation of Boolean functions. The problem with tables is that their size is exponential in the number of variables: that is, lots of memory is needed to keep the entries of the tables.

Example 1: A three-variable completely specified Boolean function $f1$, as shown in Table 5.1, uses an eight-entry truth table, even though only two combinations of the input variables produce a 1.

a	b	c	$f1(a, b, c)$
0	0	0	0
0	0	1	0
0	1	0	0
0	1	1	1
1	0	0	0
1	0	1	1
1	1	0	0
1	1	1	0

Table 5.1: The Truth Table of Boolean Function $f1$.

One obvious way to try to compact this representation is the notion of an ON-set. For a completely specified Boolean function, only those entries in the truth table that contribute to a 1 value are explicitly specified. The rest of the entries are assumed to produce 0's. Under this assumption, the ON-set of the Boolean function $f1$ is

$$f1(a,b,c) = \{(0\ 1\ 1), (1\ 0\ 1)\}$$

An obvious generalization of this specification to an incompletely specified Boolean function is the inclusion of OFF-set or DON'T CARE-set (DC-set).

Example 2: Consider the incompletely specified Boolean function $f2$ as shown in Table 5.2. The "X" denotes "Don't Care." The ON-set and DC-set of this function are $\{(0\ 1\ 1), (1\ 0\ 1)\}$ and $\{(1\ 1\ 1)\}$, respectively.

5.2.2 Sum-of-Products and Product-of-Sums Forms

Sum-of-products (SOP) form and its dual (POS) are the classical ways to represent a Boolean function in a two-level format. Sum-of-products form is a "two-level" representation because the Boolean function can be implemented

a	b	c	$f2(a,\ b,\ c)$
0	0	0	0
0	0	1	0
0	1	0	0
0	1	1	1
1	0	0	0
1	0	1	1
1	1	0	0
1	1	1	X

Table 5.2: The Truth Table of Boolean Function $f2$.

by AND gates to generate the product terms followed by an OR gate to generate the sum. For example, for the completely specified function $f1$ in Example 1, the SOP form representation is

$$f1(a,\ b,\ c) = a'bc + ab'c$$

Methods to achieve the representation of a Boolean function in its "minimal" sum-of-products form have been the subject of research for decades. The cost function to minimize can be the number of terms or the total number of *literals* (see Section 5.2.3) in the Boolean function. Among many techniques, the most commonly described are the Karnaugh map and Quine-McCluskey methods. In this chapter, the focus is on the technique to achieve the minimal representation of a Boolean function in *factored form*.

To recall the Karnaugh map minimization method, consider the incompletely specified Boolean function $f2$ as given before. The Karnaugh map technique of minimization involves representing an n-variable Boolean function as a Karnaugh map, which is an n-cube drawn in two dimensions with the nodes of the cube adjacent to each other. The next step is to find, by inspection, the largest rectangle (or a combination of rectangles) that covers all the 1's. The rectangle must contain no 0's. We can use the "X" to our advantage to maximize the size of a rectangle that covers the 1's (or 1-cells).

Example 3: The Karnaugh map of the function $f2(a, b, c)$ presented in Example 2 is

c \ ab	00	01	11	10
0				
1		1	X	1

By covering the (overlapping) rectangles $\{a'bc, abc\}$, and $\{abc, ab'c\}$, we can express $f2$ as

$$f2(a,b,c) = (a+a')bc + a(b+b')c = bc + ac = c(a+b)$$

This procedure, in effect, is simplifying the expression by making use of one of the Boolean identities:

$$a + a' = 1 \text{ (complement)} \quad (5.1)$$

$$a + ab = a \text{ (absorption)} \quad (5.2)$$

$$ab + a'c + bc = ab + a'c \text{ (consensus)} \quad (5.3)$$

We particularly draw the reader's attention to this covering technique, which will be used later in the context for the minimization of multilevel logic functions.

5.2.3 Definitions

After the informal introduction, definitions are needed to facilitate the discussion of the subject in more concrete terms.

A *variable* is a symbol representing a single coordinate of the Boolean space, for example a. The *literal count* of a variable in an expression is the number of times the variable or its negation appears in the right-hand side of the expression. The literal count of an expression is the sum of the literal counts of all its variables.

A *cube* (implicant) is a *set* C of literals such that $x \in C$ implies $x' \notin C$. A cube represents the *conjunction* of its literals (a product term). For example, $\{a, b, c'\}$ is a cube, and there are times that we would rather say abc' is a cube, for simplicity. A *prime cube*, also called a prime implicant, is a cube that is not contained by another cube. In the context of Karnaugh maps, a cube corresponds to a rectangle in the Karnaugh map. An *essential cube* (or essential implicant) is an absolutely necessary "rectangle" in the Karnaugh map for the function. An *expression* is a *set* f of cubes. An expression represents the *disjunction* of its cubes. For example, $\{\{a\}, \{b, c'\}\}$ or simply $\{a, bc'\}$, or $a + bc'$, is an expression.

- Nonredundant expression: no cube in the expression properly contains another.
- Redundant expression: for example, $\{\{a\}, \{a, b\}\}$ is redundant.

Following these definitions, a *Boolean expression* is a nonredundant expression. An *expression* provides a natural representation of the sum-of-products form of a Boolean function. The *support* of an expression is

$$sup(f) = \{x \mid \exists \; cube \; C \in f \text{ such that } x \in C \; or \; x' \in C\} \quad (5.4)$$

In other words, it is the set of variables that f depends on. The notion of "support" is vitally important in factorization and technology mapping for look-up table-based FPGAs, which will be discussed in later sections.

5.2.4 Covering

A few of the minimization problems that we shall discuss can be formulated under the general framework of *covering*. We shall first provide a formalism for covering, and illustrate with an example. Given a set $\mathcal{X}$, a collection of subsets $\mathcal{C} = \{S_i | S_i \subseteq \mathcal{X}\}$, and a cost function $cost(S_i)$, find a *cover* $\mathcal{C}' \subseteq \mathcal{C}$ such that $\bigcup_{S_i \in \mathcal{C}'}(S_i) = \mathcal{X}$ and $\sum_{S_i \in \mathcal{C}'} cost(S_i)$ is minimized.

Example 4: Consider a completely specified Boolean function represented by the following Karnaugh map:

c \ ab	00	01	11	10
0				1
1		1	1	1

Following the notation defined, we have

$$\begin{aligned}
\mathcal{X} &= \{\{ab'c'\}, \{a'bc\}, \{abc\}, \{ab'c\}\} = \text{all } \boxed{1} \text{ cells} \\
\mathcal{C} &= \text{all implicants} \\
cost(\text{implicant}) &= \text{number of literals in the implicant} \\
\mathcal{C}' &= \text{cover} = \{\{bc\}, \{ab'\}\}
\end{aligned}$$

Here the implicant $\{bc\}$ covers two $\boxed{1}$ cells $\{\{abc\}, \{a'bc\}\}$, and similarly the implicant $\{ab'\}$ covers two $\boxed{1}$ cells $\{\{ab'c\}, \{ab'c'\}\}$.

5.2.5 Factored Forms

The sum-of-products form and its dual (POS) are the classical ways to represent a Boolean function in a two-level format. For implementation of Boolean function in multilevel logic devices, we resort to the *factored form*. The factored form for representing Boolean functions is defined recursively as

1. A literal is a factored form.
2. A sum of factored forms is a factored form.
3. A product of factored forms is a factored form.

It is important to notice that factored forms are not unique.

Example 5: The expression

$$f1 = abd + acd + ab'd'$$

can be written in factored form with six literals as

$$f1 = a((b + c)d + b'd')$$

To illustrate the nonuniqueness of factored forms, for example, the same function can also be represented in factored form (with seven literals) as

$$f1 = ad(c + b) + ab'd'$$

Like the SOP form representation, we prefer to present the factored form with the least number of literals. Readers may find the factored form reminiscent of the representation of a dividend by its quotient, divisor, and residual. There are indeed some connections to that, as we shall see. The factored form can be used to represent logic expressions for efficient implementation, just as Horner's rule is used to represent an expression in parenthesized notation to reduce the total number of operations.

Exercise 1: Is the following Boolean function f in factored form?

$$f = abed' + abgd + abe'd + aced'$$

5.2.6 Factoring

Factoring is the process of representing a Boolean expression in factored form. Factoring of an expression f can be performed by choosing a *divisor* k (or h) of f to obtain

$$f = kh + r$$

and then the subexpressions k, h, and r can be recursively factored.

5.3 Multilevel Logic Minimization Methods

The primary reason why we focus on multilevel logic minimization methods as opposed to two-level is that FPGAs are multilevel logic devices. Each configurable logic block (CLB) in a look-up table-based FPGA takes a limited number of inputs, 4 or 5, and its look-up table is general purpose. Therefore, it is not advantageous to implement a Boolean function in sum-of-products form using this FPGA technology. Sum-of-products form is more appropriate for programmable logic devices that are primarily two-level networks. Multilevel networks like Xilinx FPGAs are suitable for implementation of a Boolean function using more than two levels of CLBs. More importantly, a multilevel logic network allows the *sharing* of resources for designs with multiple outputs. To implement Boolean functions as multilevel logic networks effectively, we need to understand the basic idea of factorization and decomposition. We shall put less emphasis on the simplification of Boolean functions. Rather, we focus the discussion on the presentation of Boolean functions in multiple-level form.

5.3.1 Logic Minimization Program misII

MisII (its successor is **SIS**, with sequential synthesis commands) is a research tool for multilevel logic minimization produced by Professors Brayton and Sangiovanni and coworkers of UC Berkeley. Much of the material presented in this chapter has originated from research papers written by their research group.

Keep in mind that the number of logic expressions to be minimized in real designs can be quite large. The logic minimization tool promotes the effective use of resources (for example, CLBs) by sharing them. To facilitate sharing, we need to manipulate and represent Boolean functions at the granularity that "entities" can be shared. This is the basic idea of the major operations: **gcx**, **gkx**, and **decomp** in **misII**. In these operations, **misII** attempts to break complex logic expressions into smaller subexpressions by *factorization* or *decomposition* and minimizes resource utilization by sharing common subexpressions. Readers will notice that this idea is reminiscent of the process of extracting the greatest common divisor (GCD) from two numbers by repeated division.

Exercise 2: Find the GCD between the numbers 156 and 244 by repeated division.

We illustrate the key ideas of the Boolean minimization process with the following detailed example.

Example 6: Suppose that we have the following logic network represented as two Boolean expressions in SOP form:

$$\begin{aligned} f1 &= abcd + abce + ab'cd' + ab'c'd' + ab'c'd' + a'c + cdf + abc'd'e' + ab'c'df' \\ f2 &= bdg + b'dfg + b'd'g + bg'eg \end{aligned} \tag{5.5}$$

There are obvious redundancies in the expressions – they can be *simplified* by repeated use of the identities (5.1), and (5.2). With some work, the Boolean functions can be simplified to

$$\begin{aligned} f1 &= ab'c'f' + ab'd' + ac'd'e' + a'c + bcd + bce + cdf \\ f2 &= bdg + b'd'g + dfg \end{aligned} \tag{5.6}$$

The next step is to manipulate functions to the granularity that "entities" can be shared. The factored form provides a means but not necessarily the only means (there are other forms of representations, such as binary decision diagrams, or if-then-else graphs [7]). So in their factored forms,

$$\begin{aligned} f1 &= a(b'c'f' + d'(c'e' + b')) + c(d(f + b) + be + a') \\ f2 &= g(d(f + b) + b'd') \end{aligned} \tag{5.7}$$

The final step is to extract the commonalities among the Boolean functions by decomposing (kernel decomposition) large expressions into smaller expressions and representing the functions in terms of their common subexpressions. Some intermediate variables are generated as a result of this factorization process; they are the variables in square brackets:

$$\begin{aligned} f1 &= c(be + [44] + a') + a(b'c'f' + d'(c'e' + b')) \\ f2 &= g(b'd' + [44]) \\ [44] &= d(f + b) \end{aligned} \tag{5.8}$$

There are times that this level of granularity is not desirable. The following sections provide some background on the theory of the aforementioned operations and hence a way to control the process of simplifying, factoring, and decomposing Boolean expressions.

5.3.2 Logic Simplification

The basic technique used in two-level minimization of single-output Boolean functions is to find a cube cover of the lowest cost. In `misII`, the equations are represented in SOP form, and `espresso` is used for simplification. The two-level logic minimization techniques such as the Quine-McCluskey method have been well studied (see previous chapter), but a computer-oriented method such as `espresso` is probably the most often used. The algorithms behind `espresso` are too vast to cover in one small section. We urge the reader to consult the definitive book "Logic Minimization Algorithms for VLSI Synthesis" for details [2].

5.3.3 Algebraic and Boolean Division

In `misII`, division is the mechanism by which Boolean expressions are manipulated to represent themselves in factored forms. There are two division methods: algebraic and Boolean. Algebraic (symbolic) division is also known as weak division. This is simply symbolic division without attaching semantics to the operators. That is, the process makes no reference to the semantics of the Boolean operations such as AND, OR, and NOT. Boolean division, also known as strong division, uses the semantics of the operators AND, OR, and NOT in the process. For example, Boolean identities (5.9) to (5.12)

$$a + a' = 1 \tag{5.9}$$

$$ab + ab' = a \tag{5.10}$$

$$aa' = 0 \tag{5.11}$$

$$ab + a'c + bc = ab + a'c \tag{5.12}$$

can be used in Boolean division to simplify expressions that cannot be performed otherwise by algebraic division. Some definitions are necessary to explain division. The *product of two expressions* f and g, fg, is the set

$$\{c_i \cup d_j | c_i \in f \text{ and } d_j \in g\}$$

made nonredundant using the standard Boolean operation of containment. Recall the definition of support of f as defined in (5.4), which is the set of variables that f depends on. A product of two expressions is an *algebraic product* when f and g have disjoint support, that is, no variables in common. A product of two expressions is a *Boolean product* when f and g have common support.

The *quotient* of an expression f by another expression g, f/g, is the *largest* set q of cubes such that f can be expressed as

$$f = qg + r$$

where q is the quotient and r is the remainder. Note that the notations of *quotient* and *divisor* are symmetrical to each other.

Exercise 3: Given the following set of Boolean expressions:

$$\begin{aligned} f &= ad + bcd + e \\ g &= a + bc \\ h &= a + b \end{aligned} \tag{5.13}$$

find their (algebraic and Boolean) quotients

$$\begin{aligned} f/g &= \\ f/h &= \end{aligned} \tag{5.14}$$

In the algebraic division method, we can distinguish between *algebraic divisors* that contain a single cube, called *single-cube divisors*, and ones that contain multiple cubes, called *multiple-cube divisors*. Recall that a cube is a *conjunction* of its literals.

Example 7: Given the following Boolean function,

$$f1 = ade + bde + cde + f = (a + b + c)de + f$$

de is a single-cube divisor, and $(a + b + c)$ is a multiple-cube divisor.

5.3.4 Kernels and Cokernels

After the factorization, we need to extract the commonalities among expressions. The notions of kernels and cokernels are introduced to provide an effective algorithmic way for finding subexpressions common to two or more expressions.

We say an expression is *cube-free* if no cube divides the expression *evenly*, that is, no literal appears in every cube of the expression. The *primary divisors* (quotients if you may) of an expression f are the *set* of expressions

$$D(f) = \{f/C \mid C \text{ is a cube}\}$$

The *kernels* of an expression f are the set of cube-free primary divisors, that is,

$$K(f) = \{g | g \in D(f) \text{ and } g \text{ is cube} - \text{free}\}$$

In other words, the kernels of an expression f are the cube-free primary divisors (quotients) of f. A kernel is what is left after a cube has been factored out. The cube C (divisor) used to obtain the kernel (quotient) $k = f/C$ is called the *cokernel* of k.

Example 8: Consider the Boolean function $f1$ and its factored form

$$\begin{aligned} f1 &= adf + aef + bdf + bef + cdf + cef + g \\ &= (a+b+c)(d+e)f + g \end{aligned} \tag{5.15}$$

A cokernel is df and its kernel is $(a + b + c)$.

Exercise 4: Determine which of the given divisors of the Boolean function $f1$ are kernels.

$$f1 = abcdg + abcdh + abce + abcf + abi \tag{5.16}$$

$$\begin{aligned} f1/a &= bcdg + bcdh + bce + bcf + bi \\ f1/ab &= cdg + cdh + ce + cf + i \\ f1/abcd &= g + h \end{aligned} \tag{5.17}$$

We classify kernels according to whether they can be factored further. Level 0 kernels are kernels that cannot be further factored. A kernel is of level n if it has at least one level $n - 1$ kernel but no kernels of level n or greater. For example, $(a + b)$ is a level 0 kernel, and $a(b + c) + d$ is a level 1 kernel.

Based on the notion of kernels, it has been proved that f and g have a common multiple-cube divisor (or quotient) d *if*, and *only if*, there exists $k_f \in K(f), k_g \in K(g)$ such that $d = k_f \cap k_g$. $K(f)$ and $K(g)$ are the set of kernels of f and g, respectively [3]. Informally, this states that two functions f and g have a multiple-cube common divisor if, and only if, the intersection of a kernel from f and a kernel from g has more than one cube. This provides a method for detecting common divisors, (or quotients) other than single cubes, by finding common kernels.

Example 9: Given Boolean functions $f1$ and $f2$

$$\begin{aligned} f1 &= abcdx + abcex + abcfx \\ f2 &= abcdx + abdex + abdfx \end{aligned} \tag{5.18}$$

the cubes and the respective kernels are

		cubes	*kernels*
$f1$	:	$(abcx)$	$(d+e+f)$
$f2$	:	$(abdx)$	$(c+e+f)$

(5.19)

The functions can be expressed in factored form as

$$\begin{aligned} f1 &= abcx(d+e+f) \\ f2 &= abdx(c+e+f) \end{aligned} \tag{5.20}$$

and the multiple-cube common divisor can be detected by intersecting their kernels, which gives $(e+f)$ (this corresponds to the **gkx** operation in `misII`). The functions can be represented by their multiple-cube common divisor as

$$\begin{aligned} f1 &= [13]abxc \\ f2 &= [14]abxd \\ [12] &= e+f \\ [13] &= [12]+d \\ [14] &= [12]+c \end{aligned} \tag{5.21}$$

Notice that we have not expressed these functions in their most simplified factored forms yet, since $f1$ and $f2$ still have the cube abx in common. Extracting the multiple-cube divisors from expressions is just one way to find subexpressions; extracting common cubes is another possibility (this corresponds to the **gcx** operation in `misII`). In `misII`, factors can be extracted by alternating **gkx** and **gcx** steps with diminishing thresholds (see Section 5.6.2).

Continuing with the previous example, we extract the common cube abx and re-express the functions in terms of this cube.

Example 10:

$$\begin{aligned} f1 &= [20][22]c \\ f2 &= [21][22]d \\ [19] &= e+f \\ [20] &= [19]+d \\ [21] &= [19]+c \\ [22] &= abx \end{aligned} \tag{5.22}$$

5.3.5 Decomposition

Disjoint Decomposition

Decomposition refers to the process according to which an n-variable function is expressed as a composition of several functions each depending on fewer than n variables. The concepts are illustrated in Fig. 5.1. In these examples, the input variables are x, y and z, and g, h, and F are combinational logic blocks. The functions f_1 and f_2 represent *disjoint decompositions*, whereas f_3 is an example of *nondisjoint decomposition*.

$$f_1(x,y) = F(g(x),y)$$

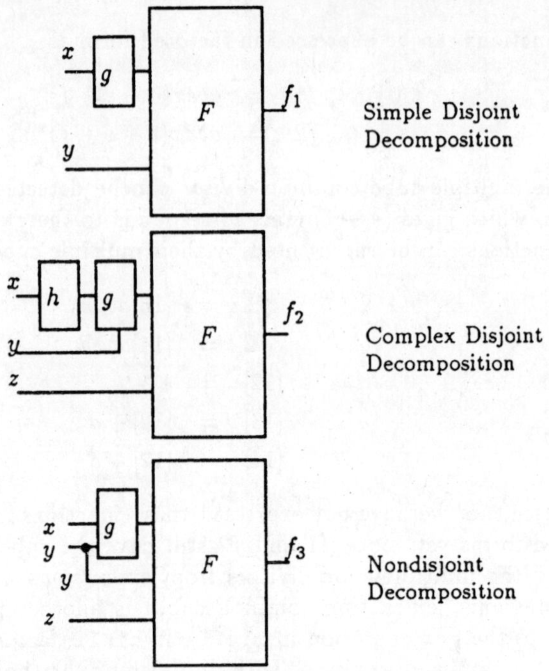

Figure 5.1: Concept of Decomposition.

$$\begin{aligned} f_2(x,y,z) &= F(g(h(x),y),z) \\ f_3(x,y,z) &= F(g(x,y),y,z) \end{aligned} \tag{5.23}$$

The decomposition schemes can be characterized by the nature of the subnetworks. f_1 is an example of *simple disjoint* decomposition because there is only one g subnetwork. While f_2 is an example of *complex disjoint* decomposition that has two subnetworks, g and h, and f_3 is a *nondisjoint decomposition* because the variable y is used as input to two blocks, g and F. Techniques used in decomposing f_1 can be used iteratively to decompose f_2.

The (nontrivial) simple disjoint decomposition of an n-variable function requires the identification of two disjoint sets of variables X_p and X_s in such a way that $f(X) = F(g(X_p), X_s)$, where s and p are the cardinalities of the two sets, respectively, and where $1 < s < n-1$, $s+p=n$, $X = X_p \cup X_s$, and $X_p \cap X_s = \emptyset$. Simple disjoint decompositions are trivial if the functions are associative. For example,

$$\begin{aligned} f_1(a,b,c,d,e,f) &= a+b+c+d+e+f \\ &= (a+b+c+d)+(e+f) \\ &= F(g(a,b,c,d),e,f) \end{aligned}$$

For functions that are not necessarily associative, a simple disjoint decomposition is possible if the residues of the function with respect to the variables in set X_s yield the following functions [1, 5]:

$$0, 1, g(X_p), g'(X_p)$$

The decomposition thus requires calculation of *all* residues of the functions for all combinations of the set of variables X_s. There are generally $2^n - n - 2$ residues for an n-variable function, thus it is computationally expensive to compute the best decomposition. Alternatively, we can use decomposition charts to determine the decomposition, by visual inspection, without calculating directly any residues. This is illustrated in the next section.

Roth-Karp and Ashenhurst Disjoint Decomposition

Roth-Karp disjoint decomposition scheme described in this section corresponds roughly to the `xl_k_decomp` (Xilinx-k-input-decomposition) operation in `misII`. This operation is suitable for look-up table-based FPGAs with k-input look-up tables [9]. Unfortunately, it is also computationally very expensive, so it is useful only for small designs with a high degree of symmetry. The decomposition scheme can be represented as

$$f(X_1, X_2, ..., X_n) = G(\alpha_1(X_L), \alpha_2(X_L), ..., \alpha_t(X_L), X_R)$$

where the set of input variables X_L and X_R is disjoint ($X_L \bigcap X_R = \emptyset$) and $X_L \bigcup X_R = \{X_1, X_2, ..., X_n\}$.

A method similar to the Roth-Karp decomposition is the Ashenhurst decomposition [1], and will be presented here to illustrate the idea. An Ashenhurst decomposition chart is an aid (very much like the Karnaugh map) to visualize the effect of the choice of residues and control variables for a function to be decomposed. For example, consider a Boolean function f with minterms [6]:

$$f(W, X, Y, Z) = \sum(0, 1, 3, 8, 9, 11, 12, 13)$$

A decomposition chart using XY as the control variables and WZ as the residue variables is shown in Fig. 5.2. The minterms of f are encircled. The control variables are XY, and the residues are WZ. From Fig. 5.2, the residues functions can be simplified to

$$\begin{aligned} R_0 &= 1 \ ; \ R_1 = Z \\ R_2 &= W \ ; \ R_3 = 0 \end{aligned}$$

and the decomposition is

$$\begin{aligned} f(W, X, Y, Z) &= X'Y'(1) + X'Y(Z) + XY'(W) + XY(0) \\ &= X'Y' + X'YZ + XY'W \\ &= G(\alpha_1(W, Z), \alpha_2(W, Z), \{X, Y\}) \end{aligned}$$

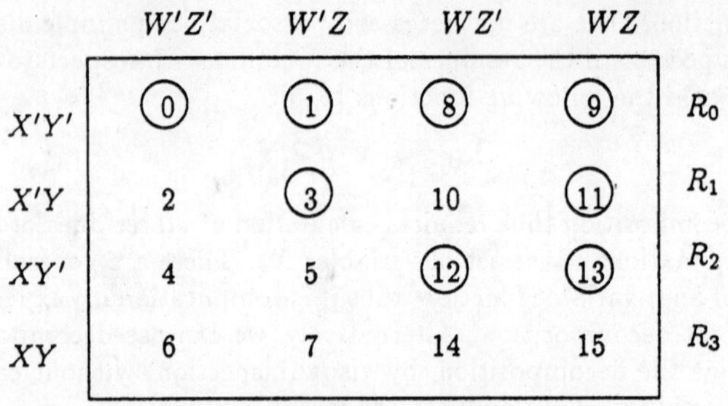

Figure 5.2: An Ashenhurst Decomposition Chart of f.

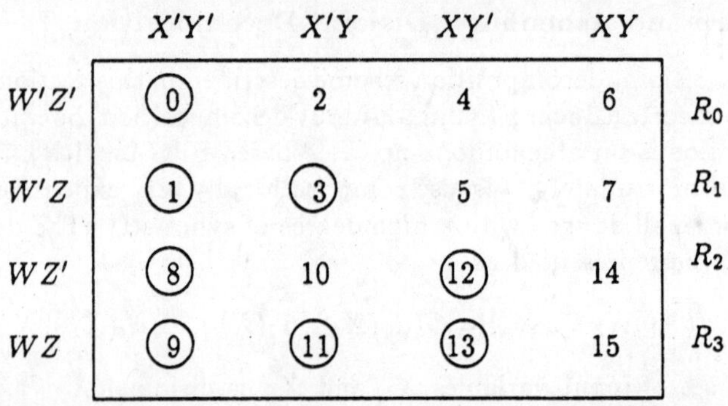

Figure 5.3: Another Ashenhurst Decomposition Chart of f.

where $\alpha_1() = W$ and $\alpha_2() = Z$. Note that we have decomposed f from a function of four-input variables (W, X, Y, Z) to a function G that depends on two sub-functions α_1 and α_2 and two control inputs X, Y. G can be implemented with a 4-to-1 multiplexer.

Notice that there can be many of these charts, depending on the choice of the combination of variables. A slightly different combination, as shown in Fig. 5.3, yields a different set of residue functions.

$$R_0 = X'Y' \quad ; \quad R_1 = X'$$
$$R_2 = Y' \quad ; \quad R_3 = (XY)'$$

The quality of decompositions should be measured with respect to the complexity of the residue functions that are yielded.

Kernel Decomposition

Kernel decomposition is the process in which all the nodes in the network are factored systematically. In kernel decomposition, factorization is the means to achieve decomposition. In decomposing a network, a node is factorized and its divisor forms as a new node in the network. The rest of the nodes in the network are then re-expressed in terms of this newly introduced node. Naturally, decomposition breaks down larger expressions into smaller pieces.

Example 11: Consider the expressions previously used in Example 9,

$$\begin{aligned} f1 &= abcdx + abcex + abcfx \\ f2 &= abcdx + abdex + abdfx \end{aligned} \tag{5.24}$$

which can be expressed in factored form as

$$\begin{aligned} f1 &= abcx(d + e + f) \\ f2 &= abdx(c + e + f) \end{aligned} \tag{5.25}$$

Decomposition expresses $f1$ and $f2$ in terms of their divisors (kernels): $d + e + f$ and $c + e + f$, respectively. The divisors form new nodes in the network, and the rest of the nodes in the network are then re-expressed in terms of these newly introduced nodes.

$$\begin{aligned} f1 &= [25][24]c \\ f2 &= [26][24]d \\ [24] &= abx \\ [25] &= d + e + f \\ [26] &= c + e + f \end{aligned} \tag{5.26}$$

5.3.6 Rectangle Covering

The reader can skip this section if he or she is not interested in the data structure used in `misII`. In `misII` the kernel intersection, the single-cube divisor extraction, and multiple-cube divisor (kernel) extraction operations are formulated and implemented by a "rectangle covering" algorithm.

The rectangle covering problem can be informally described as forming a Boolean matrix of some quantities as columns and other quantities as rows. Given the Boolean matrix, find rectangles of 1's that cover all the 1's in the matrix. The reader will find that this problem is quite similar to the Karnaugh map representation and its minimization.

Example 12: Given

$$f1 = abcdg + abceg + f$$

a cube-literal Boolean matrix can be formed, as shown in Table 5.3. Each literal occupies a column, and each cube occupies a row. For each cube, we place a 1 in the Boolean matrix whenever a literal appears in the cube.

The largest cokernel can be extracted by identifying the largest rectangle $abcg$ in the Boolean matrix. The kernel $d + e$ is indicated by the terms outside

	a	b	c	d	e	f	g
$abcdg$	1	1	1	1			1
$abceg$	1	1	1		1		1
f						1	

Table 5.3: A Cube-Literal Boolean Matrix.

the rectangle that occupy the same rows as the rectangle. Common cubes among multiple expressions can be extracted in a similar manner by building a cube-literal matrix of all the expressions.

Recall that the common multiple-cube divisor can be extracted from expressions by kernel intersection. The following example illustrates the extraction of multiple-cube divisors by building a Boolean matrix with cubes in the kernels at columns and cokernel at rows.

Example 13: Given

$$\begin{aligned} f1 &= abcdg + abceg + f \\ f2 &= abd + abe + abf \end{aligned} \quad (5.27)$$

a kernel-cokernel Boolean matrix can be formed, as shown in Table 5.4.

	d	e	f
$abcg$	1	1	
ab	1	1	1

Table 5.4: A Kernel-Cokernel Boolean Matrix.

We extract the common multiple-cube divisor by identifying the largest rectangle in the Boolean matrix, in this case, $d + e$.

Boolean matrices are usually sparse, so sparse matrix techniques are applied in the implementation for the identification of rectangles.

5.4 Technology Mapping

The logic functions generated by a technology-independent logic minimizer may not be in readily implementable form. *Technology mapping* is the process of binding technology-dependent circuits (of the target technology) to technology-independent circuits.

Logic minimizers typically produce quite fine-grained Boolean expressions. A technology mapper often groups fine-grained circuits into larger-grain cells.

But there are exceptions; for instance, the Boolean functions generated by `misII` might not be all *feasible* for Xilinx XC3000 FPGA implementation (that is, the support of some functions may be greater than 5). It requires further effort to decompose infeasible functions into feasible ones.

In the context of Xilinx FPGAs, the mapper that comes with the toolset is `xnfmap` for the XC3000 series. For the XC4000 series, the mapper, partitioner, placement, and routing tools are integrated in a program called `ppr`.

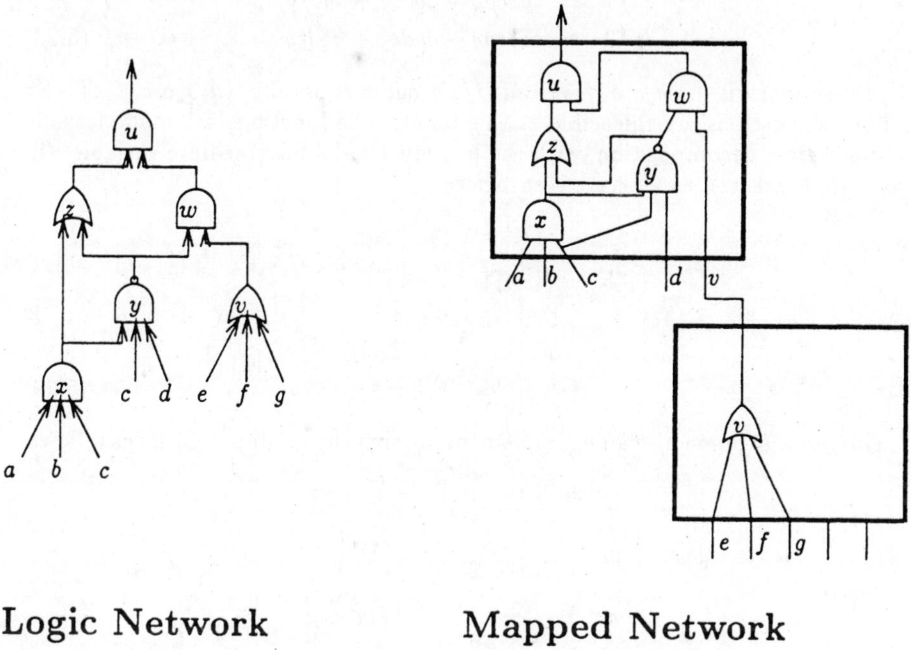

Figure 5.4: Technology Mapping of a Boolean Network Using 5-input Look-up Tables.

The technology-dependent circuits in a Xilinx FPGA are the configurable logic blocks (CLBs). They comprise look-up tables (LUTs) and flip-flops. Further, each CLB in the XC3000 series FPGA can accommodate two four-input LUTs providing the total number of distinguished inputs does not exceed five.

As we mentioned, multilevel logic networks promote the sharing of resources for designs with multiple outputs. An FPGA technology mapper for combinational circuit performs three primary functions.

1. It decomposes infeasible expressions into feasible ones.
2. It groups or pairs small expressions into CLBs to promote sharing of resources.
3. It allocates CLBs to expressions that cannot be shared.

The process of technology mapping can also be described in terms of covering. Imagine that the CLBs are templates and we are using the templates to cover the given set of Boolean expressions to be technology mapped. Figure 5.4 illustrates the technology mapping of a logic network to its mapped network implemented in look-up tables.

Example 14: Consider the following set of Boolean expressions that we used in Example 11, and that we want to implement the expressions with 5-input CLBs.

$$\begin{aligned} f1 &= abcdg + abceg + abcfg \\ f2 &= abcdg + abdeg + abdfg \end{aligned} \tag{5.28}$$

$f1$'s support is $\{a, b, c, e, f, g\}$, and $f2$'s support is also $\{a, b, c, e, f, g\}$. So both expressions are infeasible at this point. The functions are made feasible after kernel decomposition, with the introduction of intermediate variables (in square brackets), as we have seen before:

$$\begin{aligned} f1 &= [25][24]c \\ f2 &= [26][24]d \\ [24] &= abg \\ [25] &= d + e + f \\ [26] &= c + e + f \end{aligned} \tag{5.29}$$

This set of expressions can be implemented using three 5-input CLBs as follows:

$$\begin{aligned} &\#\text{CLB} \quad 1 \\ &\quad f1 = c(abg)[25] \\ &\#\text{CLB} \quad 2 \\ &\quad f2 = d(abg)[26] \\ &\#\text{CLB} \quad 3 \\ &\quad [25] = d + e + f \\ &\quad [26] = c + e + f \end{aligned} \tag{5.30}$$

CLB1 and CLB2 are single-output CLBs, and CLB3 is a two-output CLB. Note that the intermediate expressions share one CLB because they have two variables e and f in common.

5.5 Relating Literal Count to Number of CLBs

So far, we have assumed implicitly that minimizing the literal count of a set of Boolean expressions will produce a minimum number of CLBs after technology mapping. Schlag, Chan, and Kong [10] have observed an empirical relationship between literal counts (of a set of logic expressions) produced by `misII` and the number of routed CLBs produced by Xilinx technology mapper `xnfmap` with the XC3000 FPGAs.

$$\text{Number of CLBs} \approx \frac{\text{Literal Count}}{5} \tag{5.31}$$

This empirical relationship is often accurate within 20% and certainly is exact for the set of expressions presented in (5.29). But there are exceptions, as illustrated with the following example.

Example 15: Consider the following output from the high-level language **bdsyn** that describes the combinational logic of a 4-bit up-down counter (please refer to the **bdsyn** description of the counter in Chapter 3). The D's are the outputs, and up is the up/down control bit.

$$
\begin{aligned}
D0 &= [22]' \\
D1 &= [20]' \\
D2 &= [18]' \\
D3 &= [16]' \\
[0] &= Q0'\ Q1'\ Q2'\ Q3'\ up' \\
[1] &= Q0\ Q1\ Q2\ Q3'\ up \\
[2] &= Q0\ Q1\ Q2'\ up \\
[3] &= Q0'\ Q1'\ Q2'\ up' \\
[4] &= Q2'\ Q3\ up \\
[5] &= Q1'\ Q2\ Q3 \\
[6] &= Q0\ Q3\ up' \\
[7] &= Q0\ Q2\ up' \\
[8] &= Q1'\ Q2\ up \\
[9] &= Q0\ Q1\ up' \\
[10] &= Q0\ Q1'\ up \\
[11] &= Q0'\ Q1\ Q3 \\
[12] &= Q0'\ Q1\ Q2 \\
[13] &= Q0'\ Q1'\ up' \\
[14] &= Q0'\ Q1\ up \\
[15] &= Q0' \\
[16] &= [0]'[11]'[1]'[4]'[5]'[6]' \\
[18] &= [12]'[2]'[3]'[7]'[8]' \\
[20] &= [10]'[13]'[14]'[9]' \\
[22] &= [15]'
\end{aligned}
\tag{5.32}
$$

After being minimized by `misII` using the standard script, it yields with 33 literals;

$$D3 = Q3'(Q2\ D2'\ D1' + Q2'\ D2\ D1) \tag{5.33}$$

$$
\begin{aligned}
& +Q3(Q2'\ up + Q2\ Q1' + [56]) \\
D2 &= Q2'\ [70]' + Q2\ [70] \\
D1 &= [56](Q1\ Q0 + up) + [56]'(Q1'\ Q0 + up') \\
D0 &= Q0' \\
[56] &= Q0\ up' + Q1\ Q0' \\
[70] &= Q1'\ up + [56]
\end{aligned}
\tag{5.34}
$$

This might use up 4 CLBs. But the following (*collapsed*, representing in two-level format using the `collapse` command in `misII`) representation with 43 literals (in factored form) may be a better idea:

$$
\begin{aligned}
D3 &= Q0\ Q1\ Q2\ Q3'\ up + Q0\ Q3\ up' + Q0'\ Q1'\ Q2'\ Q3'\ up' + Q0'\ Q3\ up \\
&\quad + Q1\ Q3\ up' + Q1'\ Q3\ up + Q2\ Q3\ up' + Q2'\ Q3\ up \\
D2 &= Q0\ Q1\ Q2'\ up + Q0\ Q2\ up' + Q0'\ Q1'\ Q2'\ up' + Q0'\ Q2\ up \\
&\quad + Q1\ Q2\ up' + Q1'\ Q2\ up \\
D1 &= Q0\ Q1\ up' + Q0\ Q1'\ up + Q0'\ Q1\ up + Q0'\ Q1'\ up' \\
D0 &= Q0'
\end{aligned}
\tag{5.35}
$$

The technology mapping of equations (5.35) uses only two and a half 5-input CLBs.

Exercise 5: Determine the mapping of Boolean equations in (5.35) into XC3000 series CLBs. Show how the CLBs are being utilized.

During the design process prior to detailed implementation, a designer should constantly evaluate the number of basic cells (CLBs or modules) needed to implement the design. If you have a stand-alone technology mapper, then this is a simple task to do. Recent FPGA implementation tools are more integrated, and there may not be a separate technology mapper. In this case, you may need to set the appropriate flags in the integrated tools to perform technology mapping to evaluate the feasibility for implementation of the design, and skipping the place and route.

5.6 Practicalities

5.6.1 Using misII

The simplest (but blind) way of using `misII` to minimize a set of Boolean expressions contained in a file, say, `design.eqn`, is

```
%misII
misII> read_eqn design.eqn
misII> source script
```

```
misII> ps
misII> write_eqn design_sim.eqn
misII> quit
```

This treats misII like a black box without knowing what was going on, which is not a good idea and is not sufficient to produce what you would like under constraints. The misII script is a sequence of operations to minimize the set of Boolean expressions. What follows is a description of the operations included by the misII standard script.

5.6.2 MisII standard scripts

The misII package comes with three "standard" scripts.[2] The first one is the algebraic script, the second is the Boolean script, and the third one is the espresso script. They are called **script**, **script.boolean**, and **script.espressc** in the current UC Berkeley OCT 5.1 distribution.

What follows is a description of the content of the algebraic script. Keep in mind that each Boolean expression corresponds to a node in the Boolean network, so the terms "node" and "expression" are synonymous with each other. The algebraic script begins with

```
misII> sweep
misII> eliminate 5
```

These two operations partially collapse the network. The operation **sweep** eliminates any buffer ($x = y$) or inverter nodes ($x = y'$) by pushing them into their fanouts. Intermediate nodes with no fanout are also deleted. The operation **eliminate 5** eliminates any nodes with a "value" less than or equal to 5 by collapsing the nodes into their fanouts. *Value* of a node is the number of literals saved in the factored form representation by leaving the node in the network (this command is relatively expensive if there are many large nodes, as their factored form will have to be calculated many times). The next step is to minimize each individual node.

```
misII> simplify -m nocomp -d
```

There is a misII alias **siml** that performs this operation. The **simplify** operation has a few command line options. For example, the **-m nocomp** option uses the **espresso** minimization method, and the **-d** option requests that the minimization does not use any Don't Cares. Here is the main difference between the algebraic script and the Boolean script in which a full **espresso** minimization with Don't Cares is used.

```
misII> resub -a
```

[2] This discussion of the misII script is originated from Karen Bartlett of the University of Washington, Seattle.

Next, **resub** checks if a node (or its complement) is an algebraic divisor of another node (if the function for node **x** is a divisor of node **y**, its equation can be factored out of **y**). The **-a** option performs the operation on *all* nodes.

The next operation in the script is to generate multicube divisors with *merit* greater than or equal to 30 (-t 30). The "30" refers to 30 literals in the sum-of-products form.

```
misII> gkx -abt 30
```

All kernels are used during kernel intersection (**a** option), and the best (**b** option) of the kernel intersection is chosen. The **a** and **b** options are potentially expensive; the default would be to use just level 0 kernels. These are kernels that cannot be further factored, for example, $(a+b)$ is a level 0 kernel, $[a(b+c)+d]$ is a level 1 kernel. Both options use a heuristic for picking the intersecting kernels. If no threshold option were specified, all kernels would be generated (and substituted into the network).

Then it does a bit of cleaning up with **resub** and **sweep**.

```
misII> resub -a; sweep
```

As discussed in Section 5.3.4, extracting multicube divisors is only one way to factor the expressions. The alternative is to extract single-cube divisors. So the next step in the script is to generate single cubes with *value* greater than or equal to 30 (-t 30). The **b** option is the expensive option that inspects every shared-cube possibility.

```
misII> gcx -bt 30
misII> resub -a; sweep
```

Repeating these **gkx** and **gcx** steps with diminishing thresholds enables detection of powerful common factors. Just doing **gkx** with no threshold and then **gcx** might select a multicube divisor of *value* 1 that eliminates a common cube with *merit* 100.

```
misII> gkx -abt 10
misII> resub -a; sweep
misII> gcx -bt 10
misII> resub -a; sweep
misII> gkx -ab
misII> resub -a; sweep
misII> gcx -b
misII> resub -a; sweep
```

Now the script eliminates very-low-merit nodes and does decomposition:

```
misII> eliminate 0
misII> decomp -g *
```

The following commands are not part of the standard script, but it is advantageous to do one more cleanup operation to conclude.

```
misII> resub -a; sweep
```

5.6.3 A misII Script for Look-up-Table-Based Technology Mapping

There are no standard scripts for look-up table-based FPGA logic minimization. A technology-independent script, like the standard script, should be used for logic minimization prior to any technology-dependent minimization. The following sequence of `misII` operations perform the task of decomposing all the nodes in a set of logic expressions and prepare them for 5-input look-up table implementation.

```
misII> xl_split -n 5
misII> sweep
misII> simplify
misII> xl_partition -n 5
misII> sweep
misII> simplify
misII> xl_partition -n 5
misII> sweep
misII> xl_k_decomp -n 5
misII> sweep
```

This ensures that every node in the decomposed network has support of at most 5. For details, see [8].

Exercise 6: Use `misII` to produce the results of the simplification, factorization, and decomposition steps for the Boolean expressions as given in Example 6.

Bibliography

[1] Ashenhurst, R. L. The decomposition of switching functions. In *Proceedings of International Symposium Theory of Switching Functions*, pages 74–116, 1959.

[2] Brayton, R. K., G. D. Hachtel, C. T. McMullen, and A. L. Sangiovanni-Vincentelli. *Logic Minimization Algorithms for VLSI Synthesis.* Kluwer Academic Publishers, 1984.

[3] Brayton, R. K., and C. McMullen. The decomposition and factorization of Boolean expressions. In *Proceedings of Intl. Symp. on Circuits and Systems*, pages 49–54, Rome, May 1982.

[4] Brayton, R. K., R. Rudell, A. Sangiovanni-Vincentelli, and A. R. Wang. MIS: A Multiple-Level Logic Optimization System. *IEEE Transactions on Computer-Aided Design of Integrated Circuits and Systems*, CAD-6(6):1062–1081, Nov. 1987.

[5] Curtis, H. A. Generalized tree structure - the basic building block of an extended decomposition theory. In *Journal of the ACM*, Vol. 10, pages 562–581, 1963.

[6] Langdon, Glen G. Jr. A Decomposition Chart Technique to Aid in Realizations with Multiplexers. *IEEE Transactions on Computers*, C-27(2):157–159, Feb. 1978.

[7] Karplus, K. Using If-then-else DAGs for Multi-level Logic Minimization. In C. L. Seitz, editor, *Proceedings of the Decennial Caltech Conference on Very Large Scale Integration*, pages 101–117, Pasadena, CA, March 1989. MIT Press.

[8] Murgai, R., N. Shenoy, R. K. Brayton, and A. Sangiovanni-Vincentelli. Improved logic synthesis algorithms for table look up architectures. In *IEEE International Conference on Computer-Aided Design ICCAD-91*, pages 564–567, Santa Clara, California, November 1991.

[9] Roth, J. P., and R. Karp. Minimization Over Boolean Graphs. *IBM Journal of Research and Development*, pages 227–238, Apr. 1962.

[10] Schlag, M., P. Chan, and J. Kong. Empirical evaluation of multilevel logic minimization tools for a field programmable gate array technology. In *Proceedings of the First International Workshop on Field Programmable Logic and Applications*, pages 201–213, Sept. 1991.

Chapter 6

Finite State Machines

6.1 Introduction

We shall study the issues involving implementations of finite state machines (FSM) with FPGAs, particularly with Xilinx FPGAs. Recall that there are D-flip-flops (one in the XC2000 and two in the XC3000 and XC4000 series) inside each Configurable Logic Block (CLB) of Xilinx FPGAs. We shall implement finite state machines using D-flip-flops in a multilevel logic implementation setting. The treatment of FSM in this chapter will therefore emphasize the use of tools to design FSMs. We shall first discuss the general use of one FSM state assignment tool without regard to the underlying target technology. `Mustang` is targeted for multilevel logic minimization of the combinational part of the FSM. Later in the chapter, we shall discuss the impact of actually considering the underlying implementation technology on state assignment for FSM. Finally, we highlight the theory behind the state assignment strategies in `mustang` [1].

6.2 Finite State Machines

Finite state machines can be completely specified by their input/output descriptions, usually in the form of a state transition table or STT. The input description consists of inputs to the FSM and the FSM's current state. The output descriptions consist of outputs of the FSM and the FSM's next state. There are four major issues related to the design of an FSM.

1. Specification of the input/output description
2. Simplification of the specification to find inaccessible and indistinguishable states
3. Assigning values to the states of the FSM
4. Performance in terms of speed/area of the FSM.

We shall use a university tool **mustang** to demonstrate the design of FSMs using computer-aided design tools. Tools of similar nature that are available in the public domain include **meg**, **nova**, and **fsmc**. **Mustang** is a state assignment program. It takes a state transition table as input, assigns values to the states, and presents an *encoded* state transition table as output. You should feed the output of **mustang** (in pla format) to **misII** to minimize the combinational part of your finite state machine. You will also need the Boolean minimization program **espresso** to run **mustang**.

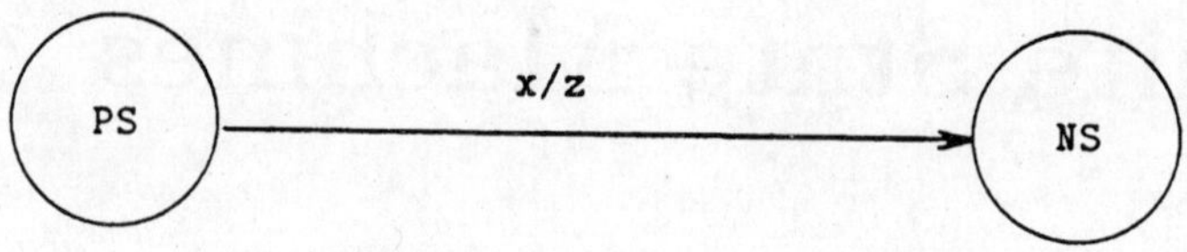

Figure 6.1: A State Transition Diagram.

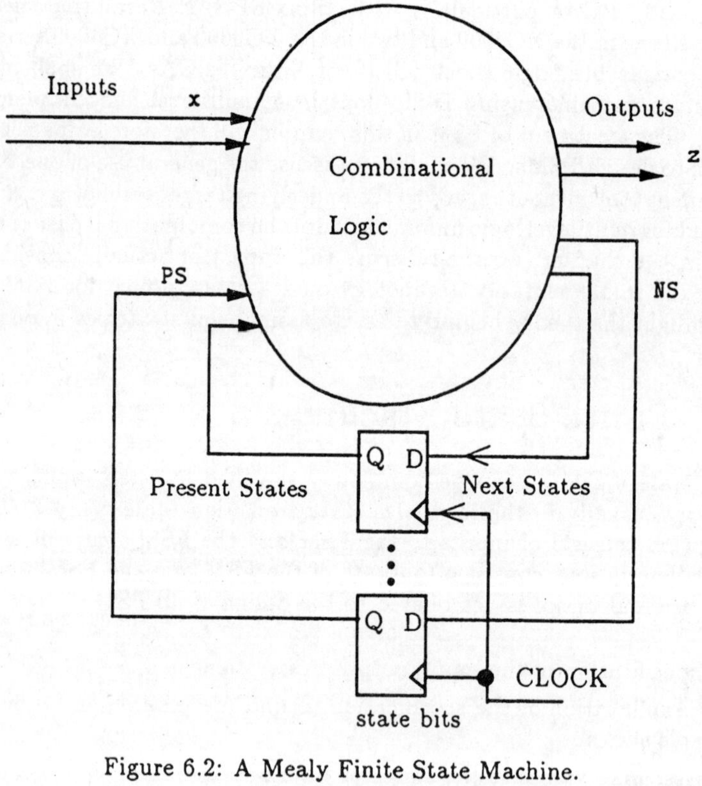

Figure 6.2: A Mealy Finite State Machine.

6.3 State Transition Table

As a shorthand notation, we symbolize `x` as the input variables, `PS` as the present state variables, `NS` as the next state variables, and `z` as the output variables. An entry in a state transition table is represented as

```
x   PS   NS   z
```

and can be depicted graphically as in Fig. 6.1.

Notice that we are adopting Mealy's notion of formulating an FSM, not Moore's notion. The subtlety of this seemingly simple description is to realize that the output `z` in Mealy's machine is produced *continuously* as a function of present state `PS` and input `x`, as illustrated in Fig. 6.2. Conceptually, it is the most convenient to think of the output `z` as being associated with the `PS`'s and input `x`. In a Moore machine, the output is as a function of the present state `PS`; the output changes only as a result of transitions from state to state.

Example 1: Consider the following three-state FSM as shown in Fig. 6.3. The FSM can also be represented in the `mustang` STT format:

```
.i 2
.o 2
.s 2
# STT format
# x PS NS  z
00  S0 S0  11
01  S0 S1  10
10  S0 S2  01
01  S1 S1  10
00  S1 S0  11
10  S2 S2  01
00  S2 S0  11
```

The first line of the description specifies the number of bits of input, the second line specifies the number of bits of output, and the third line specifies the number of bits to encode the states (not the number of states). At this point, states $S0$, $S1$ and $S2$ are abstract quantities. To implement the finite state machine as a digital system, we need to assign distinct binary values to the states. We shall consider the effect of state assignment on the cost of implementing the finite state machine. Let us pick the following state assignment:

$$\begin{aligned} S0 &= 00 \\ S1 &= 01 \\ S2 &= 10 \end{aligned} \tag{6.1}$$

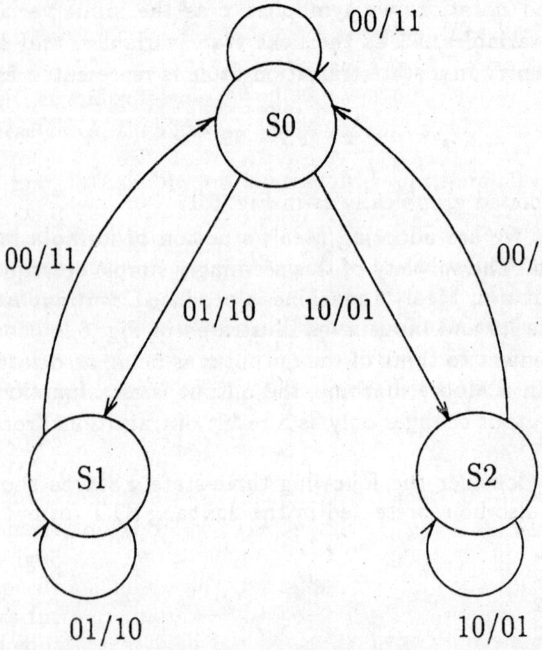

Figure 6.3: State Diagram of Example Finite State Machine.

$x1$ $x0$	$PS1$ $PS0$	$NS1$ $NS0$	$z1$ $z0$
00	00	00	11
01	00	01	10
10	00	10	01
01	01	01	10
00	01	00	11
10	10	10	01
00	10	00	11

Table 6.1: A State Transition Table.

The input/output description of the FSM is shown in Table 6.1. The combinational part of this FSM can be as simple as

$$\begin{aligned} NS1 &= PS0'\ x1\ x0' \\ NS0 &= PS1'\ x1'\ x0 \\ z1 &= x1'\ (PS0'\ x0' + PS1') \\ z0 &= x0'\ (PS0' + PS1'\ x1') \end{aligned} \tag{6.2}$$

with 14 literals, whereas a different state assignment

$$\begin{aligned} S0 &= 11 \\ S1 &= 10 \\ S2 &= 01 \end{aligned} \tag{6.3}$$

yields a slightly more complicated combinational logic of the FSM,

$$\begin{aligned} NS1 &= PS0\ x1\ x0' + [69]\ x1'\ x0 \\ z1 &= [69]\ x1' + [70] \\ z0, NS0 &= PS0\ x0' + [70] \\ [69] &= PS1'\ PS0 + PS1\ PS0' \\ [70] &= PS1\ x1'\ x0' \end{aligned} \tag{6.4}$$

with 19 literals.

For small FSMs, it is not hard to exhaust the impact of different state assignments on the complexity of the combinational logic of the FSM. This will not be practical for FSMs with more than 4 bits [2]. For the FSM that we used in Example 1, there are three distinct assignments using two bits. In general, for an n-state FSM using m bits to encode, there are

$$\frac{(2^m - 1)!}{(2^m - n)!m!}$$

possible state encoding permutations. There are CAD tools that exercise heuristics to find good but not necessarily optimal state assignments. We shall highlight the heuristics used in a state assignment program **mustang** in Section 6.6. Before we discuss the theory behind **mustang**, let us illustrate how to use or run the program under the UNIX environment. You will also need the Boolean minimization program **espresso** to run **mustang**.

Example 2: Consider the following five-state FSM `host.fsm`:

```
.i 4
.o 4
.s 5
```

```
--0- S0 S0 1001
111- S0 S0 1001
011- S0 S1 1011
101- S0 S1 1000
-1-0 S1 S1 1011
-1-1 S1 S1 0011
-00- S1 S1 1001
-01- S1 S2 1001
--1- S2 S2 1101
--0- S2 S3 1101
--0- S3 S3 1101
--1- S3 S4 1011
---0 S4 S4 1011
-0-1 S4 S4 1011
-1-1 S4 S1 0011
```

Notice that "-" represents "don't care" in the STT format. On any UNIX machine, we run **mustang** with the default settings. There are a number of flags available in the command line to control the state assignment strategies/heuristics. For example, **-ran** is a random assignment, **-l** is one-hot encoding, and **-p** uses a "fanout-oriented" algorithm to produce an encoding of states. We defer the discussion of the different encoding strategies later in Section 6.6. At this point, you simply should type

```
unix% mustang -l host.fsm  > host.pla
```

Mustang produces a **pla** file ready to be minimized by **misII**. Recall that **NS**'s are the next-state variables, **PS**'s are the present-state variables, **x**'s are the input variables, and **z**'s are the output variables.

```
.i 9
.o 9
.ilb x3 x2 x1 x0 PS4 PS3 PS2 PS1 PS0
.ob NS4 NS3 NS2 NS1 NS0 z3 z2 z1 z0
--0- 1---- 10000 1001
111- 1---- 10000 1001
011- 1---- 01000 1011
101- 1---- 01000 1000
-1-0 -1--- 01000 1011
-1-1 -1--- 01000 0011
-00- -1--- 01000 1001
-01- -1--- 00100 1001
--1- --1-- 00100 1101
--0- --1-- 00010 1101
--0- ---1- 00010 1101
--1- ---1- 00001 1011
---0 ----1 00001 1011
-0-1 ----1 00001 1011
-1-1 ----1 01000 0011
```

The following sequence of commands will guide you to minimize the logic and produce a Boolean equation file `host.eqn` that describes the combinational logic of the finite state machine. First, we feed the pla file to `misII`; then we minimize the combinational logic using the standard script. Use the *print_stats* (ps) command in `misII` to report the literal count of the minimized logic. We display the Boolean equations in their sum-of-products form, so the literal count in sum-of-products form is the same as the literal count in factored form. Next, we use the standard script one more time to minimize the logic expressions.

```
unix% > misII
UC Berkeley, MIS Release #2.0 (compiled 17-May-88 at 5:50 PM)
misII> read_pla host.pla
misII> source script
misII> ps
host.pla pi= 9 po= 9 nodes= 16 lits(sop)= 57 lits(fac)= 57
misII> pf *
   [2] = x3' x2 x1 PS4
   {NS4} = x2 [37] + x1' PS4
   {NS3} = x2' x1 [37] + PS0 [33]' + x1' PS3 + [35]
   {NS2} = x1 [38]
   {NS1} = x1' [34]
   {NS0} = PS0 [33] + x1 PS1
   {z3} = [33]' [39] + x1' PS4 + [37] + [34] + [2]
   {z2} = x1' PS1 + PS2
   {z1} = x1 PS1 + [35] + PS0
   {z0} = [34] + {NS4} + [2] + PS0 + PS3
   [33] = x0' + x2'
   [34] = PS1 + PS2
   [35] = x2 PS3 + [2]
   [37] = x3 PS4
   [38] = x2' PS3 + PS2
   [39] = PS0 + PS3
misII> source script
misII> resub; sw
misII> write_eqn host.eqn
misII> quit
```

6.3.1 Verifying the Machine

You may also verify your FSM at the logic level with the simulation facility `simulate` available in `misII`. In this example, we have 4 bits of input and 5 present-state bits, for a total of 9 combinational inputs.

```
misII> simulate 1 1 1 0 1 0 0 0 0
  1 0 0 0 0 1 0 0 1
```

This returns 5 next-state bits and 4 output bits.

6.4 State Assignment for FPGAs

In classical logic design texts, students were primarily instructed that minimum-length encodings are the best choice for assigning values to states of an FSM. Perhaps they are thought to be because minimum-length encodings use the minimum number of flip-flops. Loosely speaking, the combinational logic of a minimum-length-encoded FSM consists of two parts. The first part encodes the inputs and the present states to produce the next states. The second part decodes the present states and the inputs to produce the outputs. The encoding portion of the combinational logic typically requires higher fanin than the decoding portion.

When we assign values to the states of FSM targeted for FPGAs, we have different considerations. Each CLB of the Xilinx XC3000 FPGA has two flip-flops and one look-up tables. All the flip-flops are built into the device so it is a matter of configuring the device to use the flip-flops. Can we take advantage of the abundancy of flip-flops and use nonminimum-length encoding schemes to reduce the logic complexity of an FSM?

According to our study [3], the `-l` option of `mustang` is perhaps the best option for XC3000 series Xilinx FPGA. This is the *one-hot encoding* (also known as bit-per-state, BPS) *scheme* with the number of flip-flops used equal to the number of states. In other words, there is exactly one bit that is ON in each encoded state. The Actel FPGA design guide also advocates using one-hot encoding for FSMs targeted for Actel FPGAs. Contrary to minimum-length-encoded FSMs, the combinational logic of a one-hot-encoded FSM does not contain any encoding logic. Hence the fanin requirement of one-hot-encoded FSMs might be lower than their minimum-length-encoded counterparts. Low fanin requirement is important to any FPGA implementation, because the number of input pins to (any) FPGA basic cells are bounded.

There are obvious objections to the one-hot scheme of encoding; for example, the number of flip-flops used is certainly much higher than those minimum-length encoding schemes. Other objections according to Unger [4], are

> It is a common belief that the cost in logic complexity of one-hot encoding is usually somewhat higher than for other methods, but it is generally not far out of line. Moreover, because the transitions in one-hot encoding are all two-step, it leads to circuits slower than could be built employing a single-transition-time assignment.

To counteract these arguments, we present the logic equations of the one-hot-encoded FSM `bbara` (one of the Microelectronics Center of North Carolina, logic synthesis and optimization benchmarks) and the one generated by the minimum-length fanout-oriented option (`-tp`) in Tables 6.2.(a) and (b), respectively.

```
INORDER = x3 x2 x1 x0 PS9 PS8 PS7 PS6 PS5 PS4
      PS3 PS2 PS1 PS0
OUTORDER = NS9 NS8 NS7 NS6 NS5 NS4 NS3 NS2
      NS1 NS0 z1 z0
NS9 = PS9 [94] + [91] [95]
NS8 = [93] [96] + PS6 [91] + PS8 [86]
NS7 = [92] [97] + PS3 [91] + PS7 [86]
NS6 = PS8 [93] + PS6 [86]
NS5 = PS6 [93] + z1
NS4 = [91] [98] + PS4 [86]
NS3 = PS7 [92] + PS3 [86]
NS2 = PS3 [92] + z0
NS1 = PS4 [91] + PS1 [86]
NS0 = PS1 [91] + PS0 [86]
z1 = PS5 [84]'
z0 = PS2 [99]
[84] = x2' [86]'
[85] = PS0 + PS1 + PS4 + PS9
[86] = x0' + x1'
[91] = x3' [84]
[92] = x3 [84]
[93] = x2 [86]'
[94] = x3' x2' + [86]
[95] = PS0 + PS7 + PS8
[96] = [85] + PS2 + PS3 + PS7
[97] = [85] + PS5 + PS6 + PS8
[98] = PS2 + PS5
[99] = x3 x2' + [86]
```

(a)

```
INORDER = x3 x2 x1 x0 PS3 PS2 PS1 PS0
OUTORDER = NS3 NS2 NS1 NS0 z1 z0
[11] = PS3' PS2' [74]' [81]
[23] = PS3' PS2' PS1 PS0' [79]
[29] = PS3' [66] [79]
NS3 = [72] + z0 + [29]
NS2 = PS3' PS2 [65] [74] + PS3 [68] [80] + [73] +
      NS3 + [23]
NS1 = PS3' [86] + [68] [79] + z0 + z1 + [29] + [23] + [11]
NS0 = PS3' PS0 [90] + [81] [89] + [73] + [72]
z1 = [67] [92]
z0 = PS3 [66] [93]
[65] = x0' + x1'
[66] = PS2 PS1 PS0'
[67] = PS3' PS2 [74]'
[68] = PS2 PS1' PS0
[72] = [68] [94]
[73] = [80] [96] + z1 + [11]
[74] = PS0' + PS1'
[77] = x2' [65]'
[79] = x3 [77]
[80] = x3' [77]
[81] = x2 [65]'
[83] = x1 x0'
[84] = x1' x0
[85] = x1' x0'
[86] = PS1 [65] [88] + x3' x2' [66] + PS1' [87] + PS0 [79]
[87] = PS2' PS0 [81] + [79]
[88] = PS0' + PS2'
[89] = PS3' PS0' + [68] + [66]
[90] = PS2' [91] + PS1' [77]'
[91] = x3' x2' PS1 + [65]
[92] = [85] + [84] + [83] + [81]
[93] = [85] + [84] + [83] + [79]
[94] = PS3 [95] + PS3' [80]
[95] = [85] + [84] + [83]
[96] = PS3 [66] + [67]
```

(b)

Table 6.2: Combinational Logics of FSM `bbara` (a) One-Hot Encoded (b) `-tp` Flag Encoded.

One obvious difference between these two FSMs is the size (in literal counts) of the logic expressions of the next states variables (NS). The other difference is the number of intermediate variables. The one-hot encoding scheme is less in both counts.

Exercise 1: Use the one-hot scheme to encode the FSM in Example 1. Note the number of literals to implement the combinational logic of the FSM. Is it greater than or less than 14 literals?

6.4.1 Problem of the Initial State Assignment for One-Hot Encoding

The Xilinx XC2000 and XC3000 FPGAs clear *all* their flip-flops after configuration. This means that the value of the initial state of an FSM is all "0." But in a one-hot encoding scheme, every state has exactly one "1," so a one-hot encoded FSM will be in an illegal state after configuration. This would present a serious problem. How can we circumvent this problem?

Exercise 2: Find a general scheme to assure that a one-hot-encoded FSM (implementing with XC2000 or XC3000 FPGAs) will be in a legal state after configuration.

Exercise 3: Consider the 16-state FSM with only one input as described in Fig. 6.4. Only the values of the input are shown in the figure. Regardless of the outputs, can you suggest a good state assignment for this FSM?

6.5 Hazard and One-Hot Encoding

From a timing standpoint, a one-hot-encoded FSM has the *disadvantage* that exactly two bits are changed during each state transition. If the propagation delays of these two signals are significantly different, this would create troublesome hazards or glitches. The problem is further complicated by the fact that a designer has only limited control over technology mapping, placement, and routing during implementation. This makes propagation delays rather difficult to control. Let us illustrate the problem with an example.

Example 3: The following are the Boolean expressions of the combinational logic of a one-hot-encoded FSM.[1]

$$\begin{aligned} INORDER &= Ovflow\ CompOk\ Mode\ PS8\ PS7\ PS6\ PS5\ PS4 \\ &\quad PS3\ PS2\ PS1\ PS0; \\ OUTORDER &= NS8\ NS7\ NS6\ NS5\ NS4\ NS3\ NS2\ NS1\ NS0 \\ &\quad WE\ CS\ OE\ CountUp\ Error\ Passed; \end{aligned}$$

[1] Due to Marcelo Martin who took advanced logic design class CMPE 126 in spring 1991.

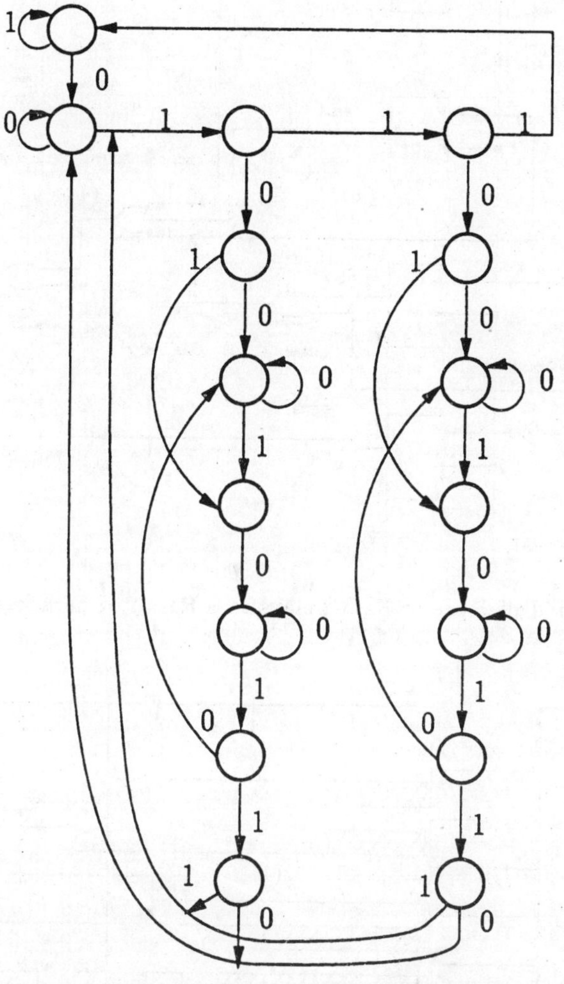

Figure 6.4: State Diagram of an Example FSM.

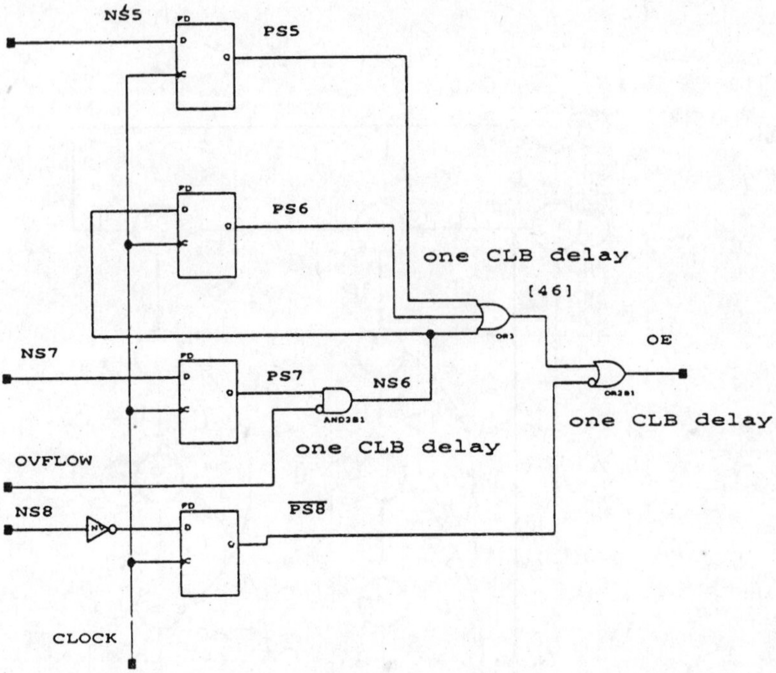

Figure 6.5: Difference in Path Delays as a Result of Technology Mapping.

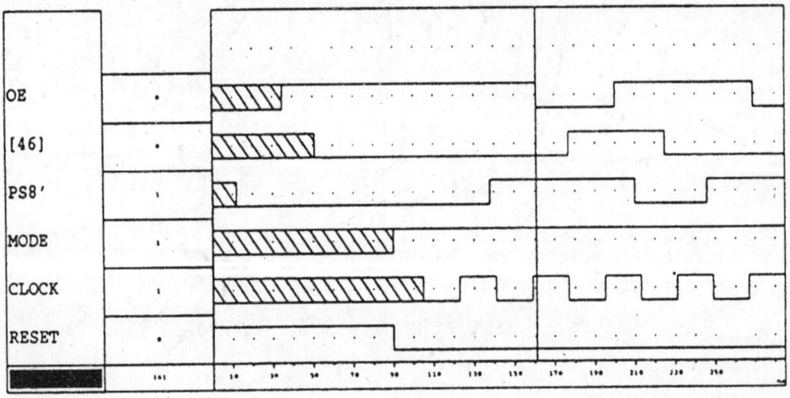

Figure 6.6: A 40 ns Glitch in OE as a Result of Differential in Path Delays.

$$\begin{aligned}
NS7 &= Ovflow'\ PS8 \\
NS5 &= Ovflow\ PS8 \\
NS8 &= PS6 \\
NS6 &= Ovflow'\ PS7 \\
NS4 &= Ovflow'\ CompOk\ PS2 + PS5 \\
NS3 &= PS4 \\
NS2 &= PS3 \\
NS1 &= [44] \\
NS0 &= [42] \\
WE &= Mode\ [46] + PS0 + PS1 + PS2 + PS3 + PS4 \\
CS &= Mode'\ [46] \\
OE &= PS8 + [46] \\
[46] &= NS6 + PS5 + PS6 \\
CountUp &= NS4 + NS6 \\
[42] &= CompOk'\ PS2 + PS0 \\
Error &= [42] \\
[44] &= Ovflow\ CompOk\ PS2 + PS1 \\
Passed &= [44]
\end{aligned}$$

Let us focus on the output OE, which is formed by

$$\begin{aligned}
OE &= PS8 + [46] \\
[46] &= NS6 + PS5 + PS6
\end{aligned} \quad (6.5)$$

Figure 6.5 depicts the logic expressions of (6.5) after being technology mapped into a XC2000 FPGA. Part of the reason for this mapping is that $NS6$ is a next-state signal and therefore has to be an "external" signal. The reset state of the FSM is $PS8 = 1$, and the FSM will transit to $PS7 = 1$ after the first clock tick. Figure 6.5 shows that the signal $PS7$ has two-CLB delays before reaching the OR2B1 gate, which generates OE, whereas the signal $PS8$ goes directly to the OR2B1 gate. The active-low signal OE is supposed to stay high. Due to the early arrival of $\overline{PS8}$ and late arrival of [46], this differential in the propagation delays generates a hazard when the FSM transits from state 7 to state 8. Considering that a XC2064PD-33 CLB has roughly a 15 ns delay and the total interconnection delay along the path from the flip-flop PS7 to OR2B1 is roughly 10 ns, this would create a 40 ns glitch in OE, as shown in Fig. 6.6. As we stated, part of the reason for this mapping is that $NS6$ in (6.5) is a next-state signal and therefore is an "external" signal by necessity. This precludes the one-CLB mapping of equations in (6.5). The problem can be circumvented partly by rearranging the expressions in (6.5) as –

$$\begin{aligned}
OE &= [99] + PS8 \\
[99] &= Ovflow'\ PS7 + PS5 + PS6
\end{aligned} \quad (6.6)$$

This allows the signal [99] to be mapped into one four-input CLB, hence reducing the differential in propagation delays between [99] and $\overline{PS8}$.

6.6 Theory of Mustang

This section can be skipped without losing much discontinuity with the rest of this book. Mustang assigns values to states of a synchronous finite state machine. It is targeted for multilevel logic implementations [1]. A state assignment is chosen –

- To heuristically minimize the number of literals in a factored form for the combinational logic of the FSM.
- To heuristically maximize the number *or* the size of the common expressions for the combinational logic of the FSM.

Mustang heuristically finds pairs or clusters of states which, if kept *minimally distant* in the Boolean space representing the encoding, result in a large number of common subexpressions in the Boolean network. It does so by attempting to maximize the number of common cubes in the logic expressions. We shall use the following example to illustrate the idea.

Example 4: The STT of a 4-state FSM is given in Table 6.3. The combina-

	Input	Present State	Next State	Output
1 >	-0	*st*0	*st*0	0
2 >	11	*st*0	*st*0	0
3 >	01	*st*0	*st*1	–
4 >	0-	*st*1	*st*1	1
5 >	11	*st*1	*st*0	0
6 >	10	*st*1	*st*2	1
7 >	1-	*st*2	*st*2	1
8 >	00	*st*2	*st*1	1
>	01	*st*2	*st*3	1
>	0-	*st*3	*st*3	1
>	11	*st*3	*st*2	1

Table 6.3: A State Transition Table.

tional logic complexity of an FSM is affected both by the state encoding and input/output specifications. First, we discuss the effect of state encoding.

Influence of Present and Next States Encoding on Logic Complexity

We shall restrict the discussion to minimum-length encoding. Let N_b be the (minimum) total number of bits to encode all the states. In Example 4, $N_b = 2$

since we only have four states. With Table 6.4, let us focus on the Present State field of the STT. Consider lines 3 and 8 in the STT, which have the same next state *st*1.

	Input	Present State	Next State	Output
3 >	01	*st*0	*st*1	–
8 >	00	*st*2	*st*1	1

Table 6.4: Part of the State Transition Table, Focusing on the Present State Field.

If we assign states *st*0 and *st*2 with codes of distance N_d, then the encoding of the state *st*1 will have a common cube with $N_b - N_d$ literals.

Let us focus on the next state field of the STT, as shown in Table 6.5. Consider lines 5 and 6 in the STT, which have the same present state *st*1. If we assign the states *st*0 and *st*2 with codes of distance N_d, then the present state *st*1 will become a common cube with $N_b - N_d$ next state bits in the encoding of *st*0 and *st*2 whatever the encoding of *st*1 is. The number of literals in the common cube will be N_b.

Influence of Inputs and Outputs on Logic Complexity

Next, we consider the effect of input and output spaces on the number of common cubes after encoding. If two different input combinations, $I1$ and $I2$, produce the same next state from different or same present states, then we have a common cube corresponding to $I1 \bigcap I2$ literals in the input space.

Encoding Strategy

So following the discussion of the influence of encoding on the number of common cubes, we understand that it is beneficial to assign adjacent values to states that are "related." The problem is that one cannot afford to try all possibilities of pairing the states. Instead, `mustang` *estimates* the gains/reduction in literal count by coding a given pair of states with close codes such that single/multiple occurrences of common cubes can be extracted.

	Input	Present State	Next State	Output
5 >	11	*st*1	*st*0	0
6 >	10	*st*1	*st*2	1

Table 6.5: Part of the State Transition Table, Focusing on the Next State Field.

The basic strategy is to build a *cost* (complete) graph $G = (V, E, W(E))$, where

- V is in one-to-one correspondence with the states
- E is a complete set of edges
- $W(E)$ are the weights associated with each edge, each weight reflecting the gains achieved by coding the states joined by the arc as closely, in terms of Boolean distance, as possible.

In `mustang`, there are two different ways of estimating $W(E)$: the first one is the fanout-oriented (`-p`) option, the other is the fanin-oriented (`-n`) option. By considering the present-state field and *output fields* in the STT, `mustang` attempts to maximize the *size* of the most frequently occurring common cubes in the encoded machine. This fanout-oriented option should be used for STTs with small input and large output spaces. By considering the next-state field and *input fields* in the STT, `mustang` attempts to maximize the *number* of occurrences of common cubes in the encoded machine. This fanin-oriented option should be used for STTs with large input and small output spaces.

Graph Embedding to Emit State Codes

After the generation of the cost graph G, `mustang` emits codes for the states by clustering states to maximize the cost. For example, suppose that we have the following complete cost graph G as shown in Fig. 6.7. We select $st3$ as the seed for clustering since the sum of any three of its edge weights is maximum. We choose any three the graph's edge weights since we have to maintain the encoding of $st3$ unidistant from the others. Given that $N_b = 3$, and we are doing a minimum-length encoding, we cluster the states

$$st3(st0, st1, st2) \tag{6.7}$$

and assign $st0, st1$, and $st2$ with unidistant code from $st3$. One possibility is

$$\begin{aligned} st3 &= 000; \qquad st0 = 001 \\ st1 &= 010; \qquad st2 = 100 \end{aligned} \tag{6.8}$$

Then we remove the node $st3$ and its edges from the cost graph. What remains is shown in Fig. 6.8. We select $st1$ as the seed for clustering since the sum of its edges is maximum. The state $st0$ can be assigned easily.

Bibliography

[1] Devadas, S., H.-K. Ma, A. R. Newton, and A. Sangiovanni-Vincentelli. MUSTANG: State assignment of finite state machines targeting multilevel logic implementations. *IEEE Transactions on Computer-Aided Design of Integrated Circuits and Systems*, CAD-7(12):1290–1299, December 1988.

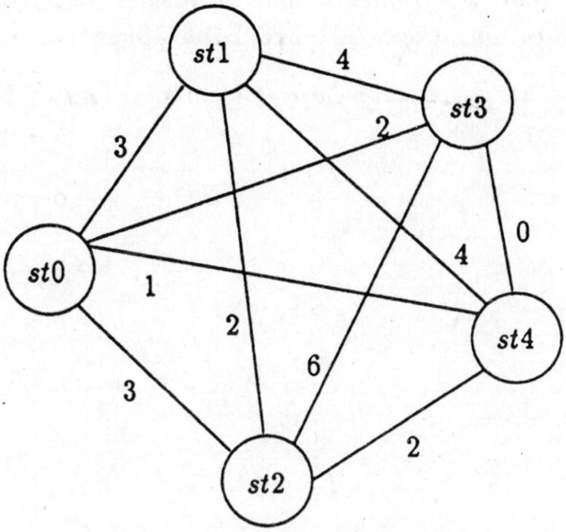

Figure 6.7: Complete Cost Graph.

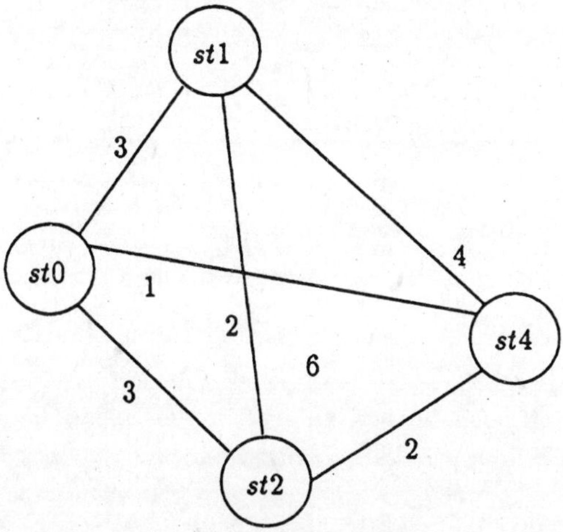

Figure 6.8: Cost Graph After Clustering of State *st3*.

[2] McCluskey, E. J. *Logic Design Principles.* Prentice Hall, Englewood Cliffs, New Jersey, 1989.

[3] Schlag, M., P. Chan, and J. Kong. Empirical evaluation of multilevel logic minimization tools for a field programmable gate array technology. In *Proceedings of the First International Workshop on Field Programmable Logic and Applications*, pages 201–213, Sept. 1991.

[4] Unger, S. H. *The Essence of Logic Circuits.* Prentice Hall, Englewood Cliffs, New Jersey, 1989.

Chapter 7

Placement and Routing

7.1 Introduction

After logic minimization and technology mapping, the design consists of a *net list* of *logical* components to be assigned to the *physical* components of the target field-programmable gate array (FPGA). Communication between the components is established by assigning the net list to wire segments and programming elements to realize the connections. This physical design process is known as *placement and routing.* Placement and routing is a crucial stage of the design flow that directly affects the feasibility and performance of the design.

Section 7.2 gives a general overview of the placement and routing process. In Section 7.3, general-purpose placement algorithms will be described. It is quite true that placement objectives and algorithms for FPGAs are variants of those that have been used for standard cells and gate array design methodology. However, routing of FPGAs departs from the traditional routing methods due to the rigid and heterogeneous natures of the FPGA routing resources. Different FPGA architectures require different routing algorithms. Routing is covered in Section 7.4.

The relation between the routing resources and the *routability* of designs is the topic of Section 7.5. An interconnect delay model commonly recognized by the FPGA manufacturers, the Elmore delay model [9], is given in the last section of the chapter. Topics in this chapter are treated at an introductory level, and references should be consulted for further details.

7.2 Placement and Routing

Placement and *routing* are two mutually dependent processes. Placement is the assignment of the logic components to particular physical components on the chip. For FPGAs, physical components such as the logic blocks are arranged in a two-dimensional array, and the I/O blocks are at the periphery of the chip. So placement in FPGAs has to deal with both the x and y directions. Interconnecting the physical components by consuming the available routing resources, wire segments and programming elements, is known as *routing.* Placement and routing of a design are traditionally carried out as two independent phases, although there are exceptions in which placement and routing are done simultaneously.

After placement, *global routing* might be performed for some FPGA place and route tools. Global routing might be viewed as a "loose" routing connecting the different logic block pins [22]. Global routing is followed by detailed routing to establish the physical interconnections by assigning programming elements along the connecting wire segments. Typically, the nets (to be routed) are ordered according to preassigned weights, and the nets are routed by the detailed router one at a time according to criticality.

Once placed and routed, a design might not function accordingly because of timing problems. Rerouting with constraints might be necessary to reduce delays in order to meet the given timing constraints. In delay-driven routing, the goal is to produce routes with reduce delays along paths that limit the overall performance of the circuit. These paths are called the *critical paths.* Delay in an FPGA circuit can be attributed to

1. Input/Output block delay
2. Logic block delay
3. Wiring delay
4. Delay due to the programming elements

Delay due to routing in FPGAs is significant: it ranges from 40% to 60% [11]. Delay calculation plays an important role in the FPGA design flow; the underlying mechanism for delay calculation is illustrated in the last section of this chapter.

As we said, placement and routing are mutually dependent. It might sometimes be necessary to create new placements to meet the design objectives: completion of routing and satisfaction of timing constraints.

7.3 Placement

In FPGA devices, component placement binds a logic unit (logic blocks, I/O blocks, tristate buffers) to a physical unit on the device. Since an FPGA has a fixed number of logic blocks, FPGA placement is not concerned with device utilization, compaction, and block orientations. Some FPGA partitioning and placement tools purposely increase utilization, using the logic blocks for routing signals, to help out the router. However, the orientations of the input and output pins of the logic block do have some bearing on the placement. For example, in the XC3000 series, the orientation of all the logic blocks is fixed. Input pins are on the west, north, and south sides of the logic block, while the output pins are on the east side. The look-up table-based logic blocks allow *some* input pins and output pins to be swapped to facilitate better placement and routing. Also, the output pins have some direct connections to neighboring logic blocks as shown in Fig. 2.6.

Placement algorithms have to confront both constraints and minimization of a *cost (objective) function*. Typical placement constraints are:

- Locking certain I/O blocks as required by the user
- Aligning tristate buffers that belong to the same bus along a long line

On the other hand,

- Minimization of the total interconnection wire lengths
- Minimization of the total number of programming elements
- Congestion reduction to ensure routability
- Minimization of connections across different regions of the chip

are common *cost functions* for gate array placement. The sole purpose is to present a good placement so that the router can complete the routing. Note that some of these cost functions may be imprecisely related to the actual cost measure. Since the actual cost measures (e.g., wire lengths) are only known after detailed routing, the placement cost function can only estimate the actual cost measure. The other consideration is the effort involved in computing the objective function efficiently and quickly. For example, a rough estimate of the wire length between two blocks a and b placed in a two-dimensional array is the Manhattan distance between the two locations:

$$|x_a - x_b| + |y_a - y_b|$$

In the Xilinx XC3000 placement and routing process, placement is done in two phases: a rough but fast *mincut-based placement* followed by a *simulated annealing-based placement*. The design implementation tool for the mincut-based placement tool is `map2lca`, and `apr` is used for detailed placement and routing.

7.3.1 Mincut-Based Placement

Real circuits usually exhibit a high degree of locality. A good placement should be able to exploit locality to cluster the tightly connected components that ought to be close together. One way to identify clusters is to partition the components into two (equal or almost equal) sets (or regions) while minimizing the number of edges crossing the two partitions. A net is said to be *cut* if it connects components in one set to components in the other. The set of nets cut is a *cutset*, and the number of nets cut is the *cutsize*. The partitioning goal is to minimize the cutsize [14, 10, 6, 2] while balancing the size of the sets.

A simplified mincut-partitioning-based placement can be illustrated as follows. The logic components of a circuit are partitioned into two regions as illustrated by the example shown in Fig. 7.1(a) [20]. The nodes a, b, c, d, e, f, and g of the circuit graph are the logic components of a circuit, and the edges are the interconnects. The logic components are then moved between the two regions, P_2 and P_1, with the objective of minimizing the number of interconnections crossing the boundary between the partitions. With some luck, when component g is selected and moved from region P_2 to P_1, the cutsize is reduced from 3 to 1, as illustrated in in Fig. 7.1(b). Each region is then partitioned again and the logic blocks arranged for minimal communication between the two regions. The partitioning process is repeated until the size of every partition is one, then we assign every logic component to a physical component on the chip.

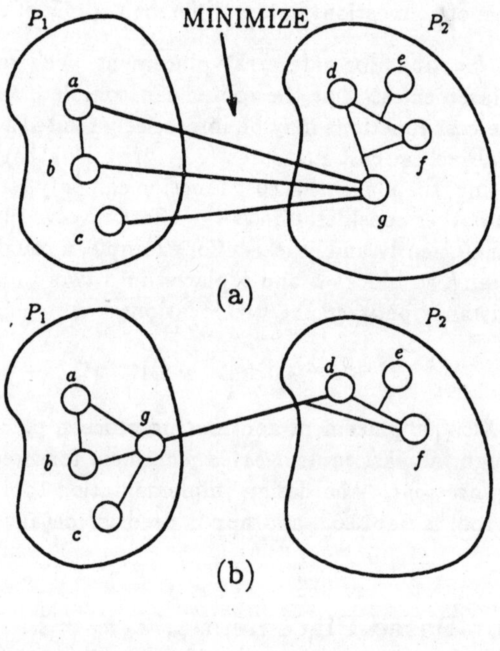

Figure 7.1: Illustration of Partitioning Based on Minimum Number of Edges Cut.

This scenario has been simplified. First, the initial partitions are generated at random, and second, we aren't always so lucky. Kernighan and Lin [14] presented the following heuristic for graph bisection while minimizing the edges cut. The basic idea of their algorithm is to start with a random partition. Then, a sequence of pairwise greedy tentative swaps of nodes (components) between the partitions is performed. The nodes tentatively swapped are those that result in the largest decrease in the cutsize *or* in the least increase. The best *partial subsequence* of the tentative swaps is then chosen as the real swaps. A new bisection is formed, and the process is repeated until the new solution is no better than the previous one. Typically, the algorithm is repeated using several different random initial partitions.

7.3.2 Iterative Improvement Placement

The placement obtained by the mincut strategy uses the number of wires crossing partitions as a metric to cluster the components. To improve the placement, a finer metric, for example, the total wire lengths, is needed. One simple technique used in iterative improvement is a *pairwise exchange* strategy that can be stated as follows. Given an initial placement,

1. Select two blocks.
2. Interchange their locations.
3. If the new placement as measured according to a *cost function* is improved, accept the interchange; otherwise, reject.
4. If too many interchanges are being rejected, stop; otherwise return to step 1.

The effectiveness of the algorithm is largely dependent on (1) the selection of the exchanged components and (2) the cost function. Comparing every component against all other components requires $(N-1)+(N-2)+\ldots+2+1 = \frac{N(N-1)}{2}$ operations, where N is the number of components available on the device. In practice, it is really not necessary to try all $\frac{N(N-1)}{2}$ exchanges, since the mincut-based placement has already identified possible clusters. It is wise to limit the number of trials by comparing with components in the *neighborhood.* If we consider a neighborhood of M components, then the number of operations will be of order, $\frac{N}{M}\frac{M(M-1)}{2} \simeq NM$only. The other factor affecting the simple exchange placement algorithm is the cost function. As mentioned, this may be the total length of the wires or the total number of programming elements.

We illustrate the pairwise interchange process with the example in Fig. 7.2. Eleven circuit components are to be placed in locations a, b, c, ..., p on a device. The top left configuration represents an initial placement of these 11 components. In applying the pairwise exchange algorithm, the total Manhattan distance is used as the cost function. The value of the cost function of the initial placement is 17. Be careful when you count the distance in Fig. 7.2, as some wires are overlapped. If f and k are exchanged (fk), the value of the cost function becomes 19. If, instead, g and k are exchanged (gk), the cost is still 19. There is no improvement for these two exchanges. Since the greedy algorithm only accepts improvements, both exchanges are rejected. The search for another pair continues.

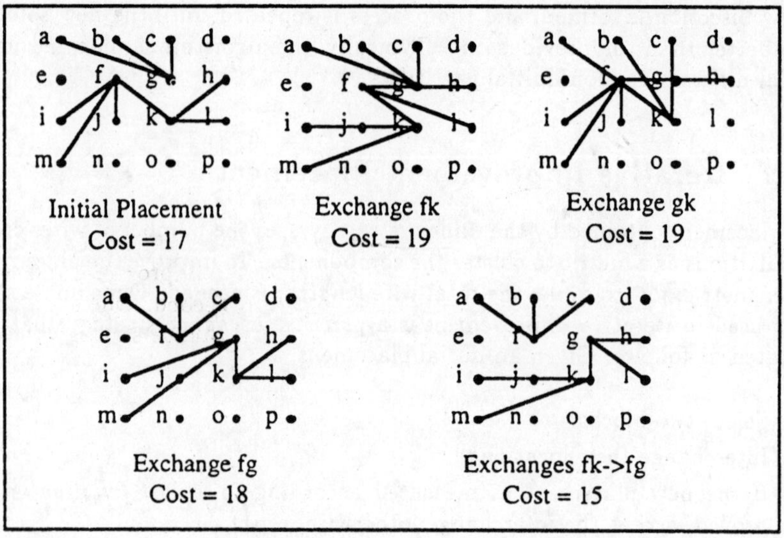

Figure 7.2: Pairwise Exchange Strategy.

7.3.3 Simulated Annealing

Instead of the greedy approach that accepts improvements only in terms of the cost (value of the cost function), it is possible to accept a slight increase in the cost that might be followed by a decrease, as it is always darkest before dawn. In Fig. 7.2 the cost is decreased after two exchanges.

Thus a temporary increase of the first exchange fk might be a *local maximum* in the cost function. In a more general fashion, the graph in Fig. 7.3 plots the cost function with respect to exchange steps. The initial cost value is indicated by point A on the graph. The cost function after the first exchange, (fk), is represented by point B. This latter point is a local maximum and a lower cost can be achieved after the next exchange, (fg), as reported by point D. The path to attain the global minimum is not unique. For example, we can start with the initial placement, exchange f and g resulting in a cost of 18 (point C on the graph), and then reach the global minimum by exchanging g and k.

There are algorithms that take into account that the *journey* to a minimum cost solution might not be monotonically decreasing. One such algorithm is called *simulated annealing* [15, 20, 22]. Kirkpatrick and coworkers applied a thermodynamically motivated simulation procedure to solve a variety of difficult optimization problems, including placement. The underlying idea is to

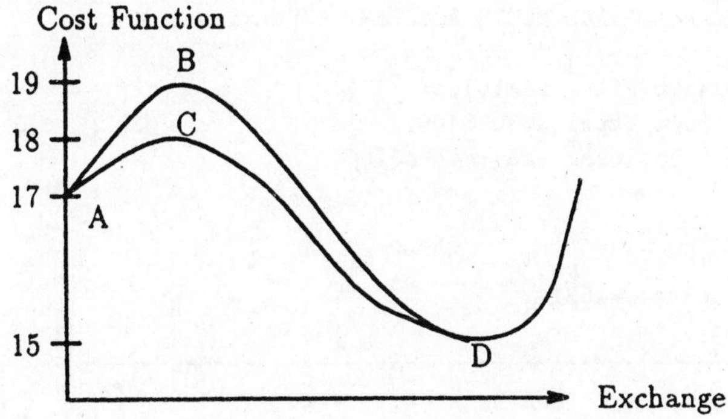

Figure 7.3: Variation in the Cost Function During Pairwise Exchange, Showing the Local Maxima and Global Minimum.

view the problem of finding feasible solutions of a combinatorial optimization problem as crystalization of a solid from its melted state at high temperature. Minimizing the cost function corresponds to minimizing the total energy of the solid by cooling down the melted solid and by performing a probabilistic state acceptance check.

We start with a randomly chosen feasible solution and a high temperature T. We perform a pairwise exchange. The *new* solution is accepted if its cost function value E_{new} is smaller than the previously accepted cost function value E_{best}. Their difference is

$$\Delta E = E_{new} - E_{best}$$

But if its cost function is worse, that is, $\Delta E \geq 0$, then we still accept the new solution with the Boltzmann probability

$$e^{-\frac{\Delta E}{T}}$$

which depends on the current temperature and the difference between the previously accepted cost function and the current value. We update E_{best} if the new solution is better. We repeat the pair exchange process at a particular temperature until there is no detectable improvement: the system is said to be at equilibrium. Then we drop the temperature gradually as the annealing progresses. At high temperatures, $e^{-\frac{\Delta E}{T}}$ is close to 1, and most exchanges are accepted. As T decreases, only small changes in the cost value are accepted. The system is said to "quench" if there is no improvement despite further reductions in temperature.

The Xilinx XC2000 and XC3000 placement tool **apr** uses this simulated annealing approach. A synopsis of a report generated by **apr** is given in Fig. 7.4 illustrating the melting, cooling, quenching, and cost (`score`) improvement during the placement process. A hint: for prototype designs where feasibility

```
AUTOMATIC PLACE AND ROUTE PROGRAM -- Version 3.30

 Input Design File: z4ml.lca
         Part Type: 3030PQ100
           Options: -a3 -r4 -s3100

Placing blocks...

Initial score: *1521*
Annealing...
--------------------------------------------------------------------
                    Best   % of    Avg                     Placement
     Temp  %Chng   Score   Init   Score   %Chng  Stdev     CPU time
--------------------------------------------------------------------
  10303.8   ****    2579   170%    5186   *****  18.6%     00:00:01
   5151.9   -50%    2579   170%    5092   -1.8%  18.9%     00:00:02
   1288.0   -50%    2579   170%    4619   -6.9%  16.5%     00:00:04
    644.0   -50%    2509   165%    4252   -7.9%  17.8%     00:00:05
    129.6   -34%    1551   102%    2648  -18.0%  16.5%     00:00:08
     96.0   -26%    1551   102%    2557   -3.4%  10.0%     00:00:09
     28.9   -18%     609    40%     816  -16.1%  14.1%     00:00:24
     11.6   -20%     468    31%     579   -7.3%   6.6%     00:00:36
      7.3   -24%     468    31%     497   -6.7%   6.3%     00:00:43
Quenching...
      0.0   ****     438    29%     438    0.0%   0.0%     00:01:13
Final score: *438*

Routing nets...

Iteration 1...
Using delay-driven detail router...
Detail Routing (with Rip-up & Retry Routing)...
Percent nets scanned: 10%.20%.30%.40%.50%.60%..70%..80%..90%..100%
Rip-up & Retry Routing and Routing Improvement
on 0 unrouted load pins in unwritten design...
Percent nets scanned: 10%.20%.30%.40%.50%.60%..70%..80%..90%..100%
There are 0 unrouted load pins.

Timing Improvement...
Using delay-driven detail router...
Percent nets scanned: 10%.20%.30%.40%.50%.60%..70%..80%..90%..100%
There are 0 unrouted load pins.
```

Figure 7.4: A Synopsis of a Report Generated by a Simulated Annealing-Based Placement and Routing Tool apr; Note the Placement Phase, Rip-up and Retry Routing Phase, and Delay-Driven Routing Phase.

or performance are not critical, forcing the simulated annealing procedure to "quench" early will deliver a fast (but not necessarily good) placement. The Xilinx XC4000 partitioning and placement tool ppr is based primarily on the mincut partitioning approach [24].

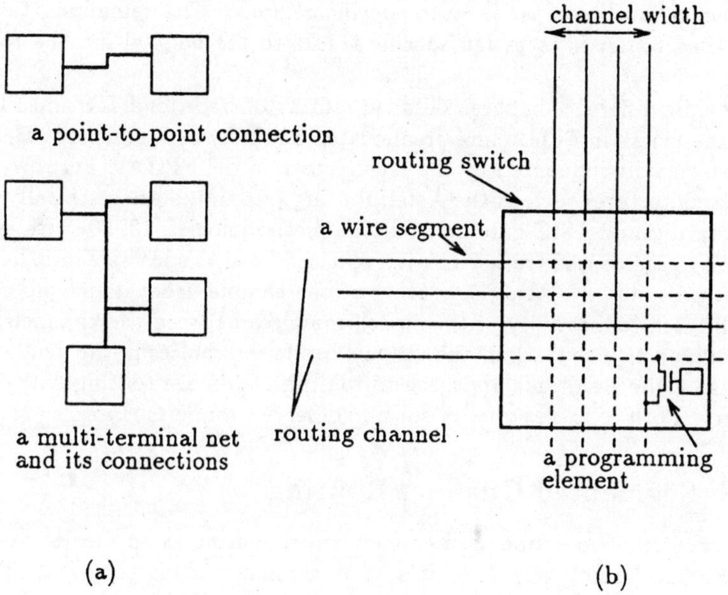

Figure 7.5: Some Routing Terminology.

7.4 Routing

The interconnections between components on integrated circuits are made by depositing metal and joining segments of metal through vias, antifuses, or RAM cells depending on the design methodology. The components may be standard cells or logic blocks in an FPGA device. The basic FPGA routing terminology is illustrated in Fig. 7.5. Generally a two-step approach to routing is employed: *global routing* followed by *detailed routing* [20].

Global routers assign "loose paths" to the nets, determining how they will navigate around the cells. These "loose paths" determine which routing areas a net will traverse to get to its destination(s). The routing areas can be channels, segments of channels, or switchboxes. For multiterminal nets, global routers must also decide in which regions the nets branch to reach different destinations. The objectives of the assignment might be to shorten the connection lengths, avoid congestion, or use fewer programming elements to decrease delay. There are many algorithms for global routing. Most of these algorithms are modifications of algorithms developed for traditional channel routing. At present, routing algorithms have been developed especially for FPGA routing primarily to reduce the delay along critical paths [11]. After the global routing phase, each net has been divided into point-to-point connections, and these connections have been assigned to specific regions. The remaining task for the detailed router is to assign specific tracks to the nets within the routing regions.

The main difference between detailed routing of traditional ICs and FPGAs is that the tracks in FPGAs are preallocated and may be segmented. Detailed routing is heavily influenced by the architecture of the FPGAs. For row-based FPGA architectures such as the Actel, routing techniques are essentially those used in traditional ASIC gate array design methodologies [13, 20] with special attention to segmented tracks. In matrix-based FPGAs such as Xilinx devices, it is not possible to isolate the routing of one channel from its neighbors due to the limited connectivity of the switch matrices between the channels. The choice of a particular track in one channel limits the choices in the neighboring channels. A more suitable approach to routing uses maze routing with rip-up and retry. Both of these types of routing deserve more discussion.

7.4.1 Segmented Channel Routing

In this section, we describe segmented channel routing as applied to the row-based FPGAs, Actel ACT 1 devices. The channel routing resources in these devices are shown in Fig. 7.6. As in the case of classical channel routing [20], segmented channel routing follows a grid of orthogonal tracks. The horizontal tracks consist of wire segments of possibly varying lengths. The vertical tracks consist of segments connected to the inputs and outputs of the logic or I/O blocks. Programmable devices, antifuses, are located between pairs of adjacent horizontal segments and at the crossing of horizontal and vertical segments.

The problem for this type of routing is as follows: given two rows of pins across a rectangular channel, connect the pins within the channel such that the connections assigned to the same track do not share segments. It also desirable to minimize congestion within the channel, as well as reduce the wire length and the number of antifuses. The interconnection paths consist of *trunks* along the horizontal tracks and *branches* along the vertical tracks. Trunks belonging to different nets should not connect. That is, there must be at least one antifuse between the segment(s) that they occupy.

This is where the segmented channel routing problem differs from the classical channel routing problem since two nets that do not overlap horizontally cannot always be placed on the same track; there must be an antifuse on the track between the two nets.

The approach to segmented channel routing depends largely on the number of tracks and their degree of segmentation. Three cases of possible segmentation are illustrated by El-Gamal in Fig. 7.7 [8]. Nonsegmented tracks imply the use of one net per track, and there is a maximum capacity (number of nets) per channel. In such an arrangement, the number of antifuses is limited (therefore less resistance), but the capacitance per track is larger than for shorter segments.

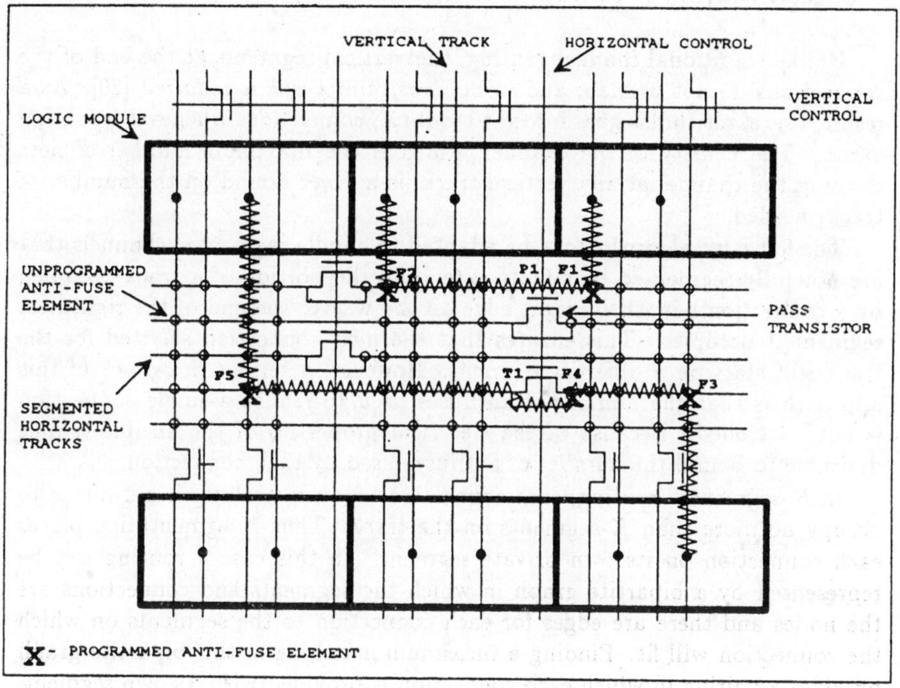

Figure 7.6: ACT1 Routing Resources.

Fully segmented tracks (segments of length one) allow maximum flexibility. As mentioned, the routing problem reduces to classical channel routing, and we can apply a commonly used algorithm, the left-edge algorithm (LEA) [12] to solve it.

Left-Edge Algorithm

Given a set of connections, C, to be laid on a set of tracks, S, where each connection spans a set vertical tracks,

1. Sort the connections in C by increasing order of their left edges (the leftmost vertical track spanned).
2. Select a track from S and assign the first connection in C to this track. Delete the track from S and delete the connection from C.
3. Assign the remaining connections as follows. Find the next connection in C whose left edge is to the right of the right edge of the last net assigned to the current track being processed. Assign connection to the current track, and delete the connection from S. If the track is now full or no more nets can be assigned to it, go back to step 2.

Unlike traditional channel routing, the vertical segments at the end of the connections do not overlap, and so no *constraint graph* is required [20]. As a result, this algorithm is guaranteed to route all connections whenever a routing exists. The density of the channel, which is the maximum number of nets crossing the channel at any vertical track, is a lower bound on the number of tracks needed.

The left-edge algorithm can be adapted to handle segmented channels that are not fully segmented by simply extending the portion of a track occupied by a connection beyond its right edge all the way to the end of the rightmost segment it occupies. This ensures that the next connection selected for the track will start on a different segment. However, a serious drawback of this approach is that the number of antifuses used to realize a single connection is not controlled. Because of the electrical properties of the antifuses, it is desirable to bound the number of antifuses used by each connection.

In K-segmented routing, each connection between two pins is permitted to occupy no more than K-segments on the track. Thus 1-segmentation places each connection on its own private segment. In this case a routing can be represented by a bipartite graph in which the segments and connections are the nodes and there are edges for each connection to the segments on which the connection will fit. Finding a maximum matching in this bipartite graph provides a pairing in which each connection is provided with its own segment. The general K-segmented routing problem is harder and, at present, requires more complicated techniques.

7.4.2 Maze Routing

Routing One Net

Most maze-routing algorithms are based on Lee's algorithm [16]. The task of a maze router is to find the shortest route between two (or more) points subject to the following restrictions:

- The route is along vertical and horizontal paths
- The routing area includes a certain configuration of obstacles that forces the router to detour.

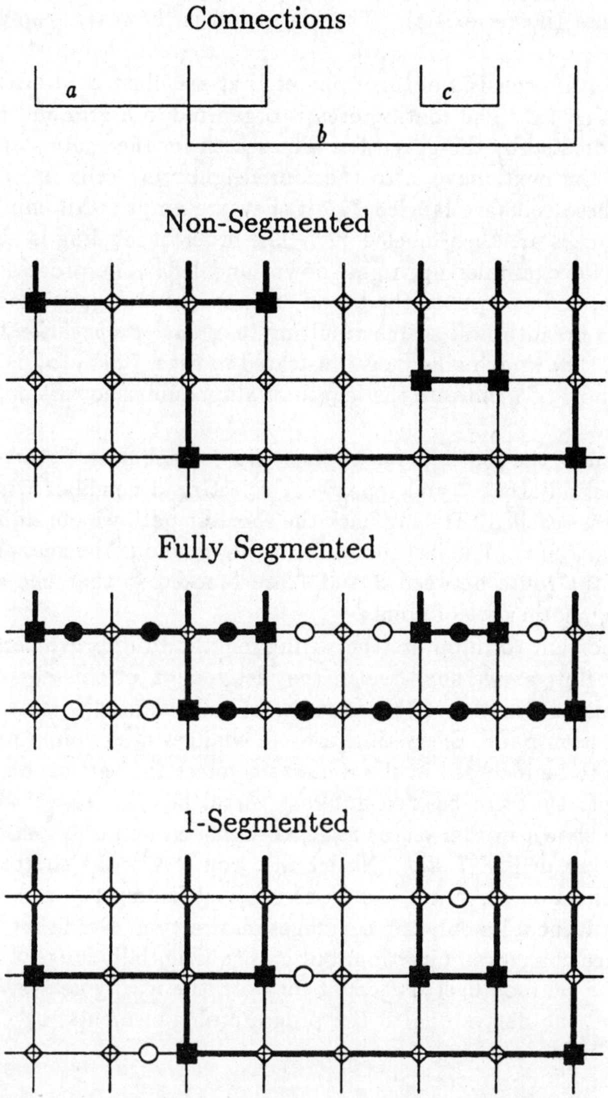

Figure 7.7: Segmented Routing. Shaded Boxes are Programmed Antifuses Between Vertical and Horizontal Tracks. Diamonds are the Unblown Antifuses. Shaded Circles are Fuses Between the Horizontal Segments. Realized Nets are in Bold Lines.

The problem may be represented as shown in Fig. 7.8(a), where S is the starting point and T is the target point. Lee's algorithm finds the shortest path between the two points (if one exists). The algorithm is, however, computationally expensive.

The algorithm consists of three phases that are illustrated briefly with the example in Fig. 7.8. The routing area is organized in a grid and includes the blockages indicated by the gray tiles. The search for the route starts at S (tile 0). From S the next move is to the four neighboring cells up, right, down, and left. These cells are labeled "1" if they are empty. All empty neighbor cells of all 1-cells are then labeled "2." The order of labeling is always in the same sense, for example: up, right, down, and left. The process is repeated for all 2-cells, and so on until the target, T, is reached. Algorithmicwise we are performing a breadth-first search resulting in a *wave propagation* from S until T is reached. The search scheme is illustrated in Figs. 7.8(b) to (d). The labels of the neighbors of T indicate the length of all possible shortest paths between the two points.

To determine the route, the next phase traces back from T starting with the smallest label adjacent T and repeatedly selecting a neighbor with a smaller label until S is reached. Tracing back the shortest path we obtain the route as shown in Fig. 7.9(a). The last phase in the algorithm is the *sweeping*: all cells included in the route between S and T are blocked so that one can proceed with routing another set of points.

It is important to minimize the wiring lengths to improve performance of the design. But, as we shall see in the last section of this chapter, routing delays are also a function of the number of programming elements included in the connection path. Every time a route changes direction, a programming element has to be included in the route to connect the vertical and horizontal wire segments that are on two different metal layers. Three routes in our example are shown in Fig. 7.9(a) as A, B, C. The number of bends along the paths are given in Fig. 7.9(b). Notice that route A is the shortest (16), but it changes direction six times. As for route B, it is only two cells longer than route A, but it includes only three changes in direction. Note that route C also has only three changes in direction, but it is substantially longer than B. Lee's algorithm can be modified to account for both the wiring delays, by counting distance, and the delays due to the programming elements, by counting the number of bends.

Routing with Rip-up and Reroute

Since a maze router routes only one net at a time, the order in which the nets are routed is important. Typically, nets are ordered according to preassigned weights and then routed one at a time by the maze router according to these weights. Due to congestion caused by previously routed nets, a router may not be able to find a path for a net to one or more of its destinations. Then, the router can enter a rip-up mode in an attempt to complete the connections

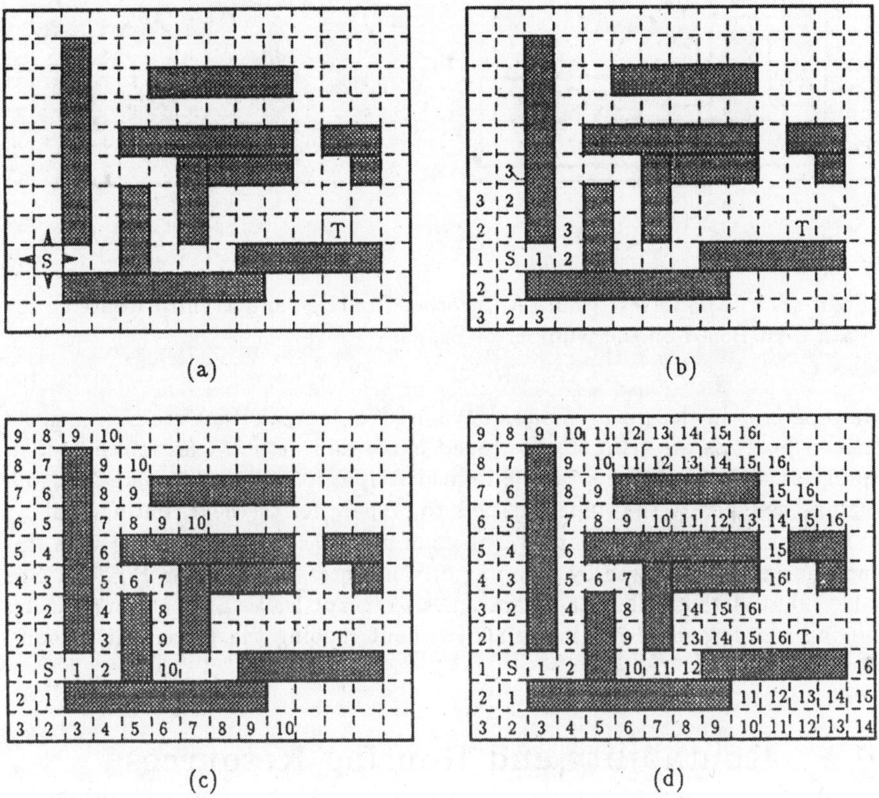

Figure 7.8: Maze Routing (a) Starting (S) and Terminal Positions (T), (b) Wave Propagation After 3 Steps, (c) After 10 Steps, (d) Eureka! Shortest Route Found.

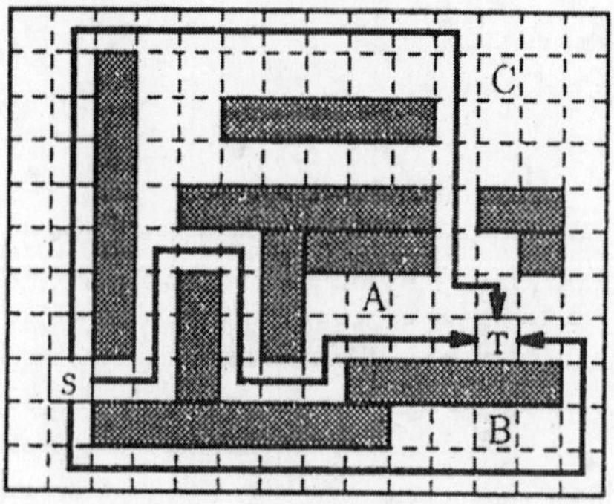

(a)

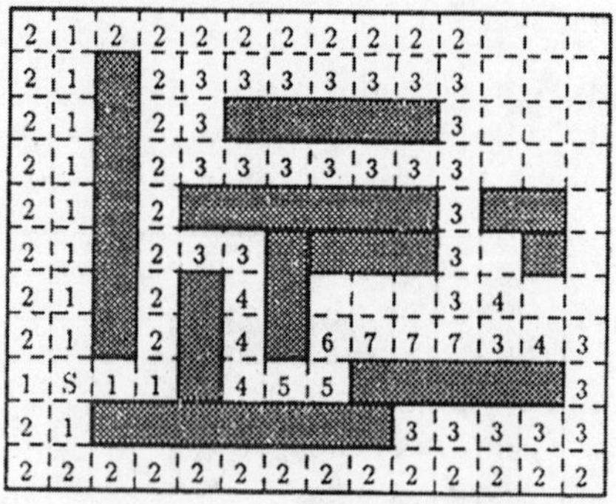

(b)

Figure 7.9: (a) Shortest Route A, Another Route B, and a Third Route C, (b) Path Cost Based on the Number of Bends.

in progress. In the rip-up mode, previously routed nets that are blocking the net in progress are temporarily ripped up to remove obstacles for the net in progress. Ripped-up nets can be immediately rerouted, and/or large weights can be assigned to the ripped-up nets to prepare for the next routing trial.

Xilinx's approach to routing consists of two phases: a maze router with rip-up and reroute followed by a timing-driven router, as shown in Fig. 7.4. See [11, 19] for details. Hint: there are times when apr leaves a few unrouted nets; giving it a few more routing trials (without redoing the placement) usually helps.

7.5 Routability and Routing Resources

7.5.1 Routability Estimation

Given an FPGA architecture, a placement and routing tool, and a design, *routability* is a measure of the *probability* that the placement and routing tool can successfully complete the routing of the design. It is the probability because almost all placement and routing tools have some degree of randomness, either in mincut-based or simulated annealing based placement. In other words, routability of a design is affected by both the routing resources available in the FPGA, and the underlying placement and routing algorithms. FPGAs have heterogeneous routing resources: direct lines, long lines, double-length wire segments. As a simplification for discussion, we shall focus only on the *widths* of the vertical and horizontal routing channels. We shall refer to them as the channel widths.

An FPGA has fixed logic and routing capacities. It is not difficult to come up with a design that is well below the logic capacity but exceeds the routing capacity. Both theory and practice have confirmed that there are three *dominant* factors that affect the routability of an FPGA design [7, 3, 4]:

- The pins per logic cell ratio, γ
- The pins per net ratio, β
- The average wire length, L

The pins-per-logic-cell ratio measures, on the average, the amount of congestion or traffic in and out of a logic cell. The pins-per-net ratio measures, on the average, the degree of branching of a multipin connection. These two ratios are characteristics of a design (before placement). Both the pins-per-logic-cell ratio and the pins-per-net ratio are available after the mapping phase of the design flow. The average wire length, on the other hand, is primarily determined by the placement and routing tools.

El-Gamal has shown that for a homogeneous two-dimensional gate array, if all the nets are point-to-point connections, that is, $\beta = 2$, then the average channel width requirement of a design is

$$W = \frac{\gamma L}{2} \tag{7.1}$$

where γ is the average number of wires emanating from each logic block and L is the average wire length [7]. For multiple-pin nets, $\beta > 2$, a possible modification of Equation (7.1) to account for multiple-pin nets is

$$W = \frac{1}{2}\left[\gamma \cdot \left(1 + \frac{\beta - 2}{\beta}\right) L\right] \tag{7.2}$$

As a good practice, to avoid futile placement and routing effort, one should check before actual place and route if the *estimated* channel width requirement of a design exceeds the available routing capacity. Both γ and β are known after the technology mapping phase of the design flow. The average wire length L can only be estimated. Typical values range from 1.0 to 2.0.

Example 1: An XC3000 design has an average pin-per-cell ratio of 6.15 and a pin-per-net ratio of 4.64. We estimate average wire length to be 1.5. Given that the XC3000 LCA architecture has five horizontal and five routing tracks, it is not likely that the design can be routed since the estimated channel width requirement is $W = 7.245$, well exceeding 5.

7.5.2 FPGA Architecture and Routability

What are the parameters in an FPGA architecture that affect routability? The answer to this question has been addressed by a recent study [21]. Brown, Rose and coworkers [3, Chap.7] have developed a theory of routability based on the probability of successful completion in routing through the connection

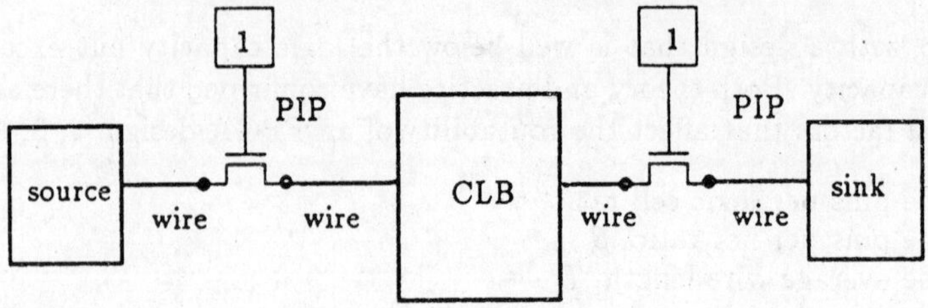

Figure 7.10: Path Delay Consists of Wiring Delays, Delay Due to the PIPs, and Logic Block Delays.

blocks and switch matrices. In their terminology, a connection block connects wire segments in the routing channels to the logic blocks. Their technique is targeted for the design of an FPGA. The design parameters are the flexibility of the connection block and the switch matrices, which have a strong bearing on the area of the FPGA chip and its routability. Their findings are reported in [3, Chap.7].

7.6 Net Delays

Different parts of a circuit contribute to path delays– the I/O pads, the logic blocks, and the interconnects as illustrated in Fig. 7.10. The source and sink components in Fig. 7.10 can be an I/O block or a logic block. For Xilinx FPGAs, the I/O block and logic block have fairly constant delays, roughly 15 ns for an I/O block and 8 ns for a combinational logic block. The delay in the CLB is due to the decoder of the RAM, which is independent of the function implemented. There is an additional delay for using a flip-flop inside the CLB.

The routing delays vary quite a bit. The interconnection delays can be attributed to delays due to the programmable interconnection points (PIPs), wiring segments, signal restoring buffers, and long lines, as depicted in Fig. 7.11. Referring to this figure, for the purpose of computing the Elmore delay [9], a PIP is modeled as a resistor of value R_{PIP}. A wire segment is modeled as a simple RC circuit with resistance R and capacitance C. For metal segments, the resistance R is relatively small in comparison to R_{PIP}, so R is often neglected in the delay calculation. The modeling of a horizontal long line with pullups at both ends and a pulldown (due to a tristate buffer) is depicted in Fig. 7.11(c). Delay calculation using the Elmore delay model will be illustrated in the next section.

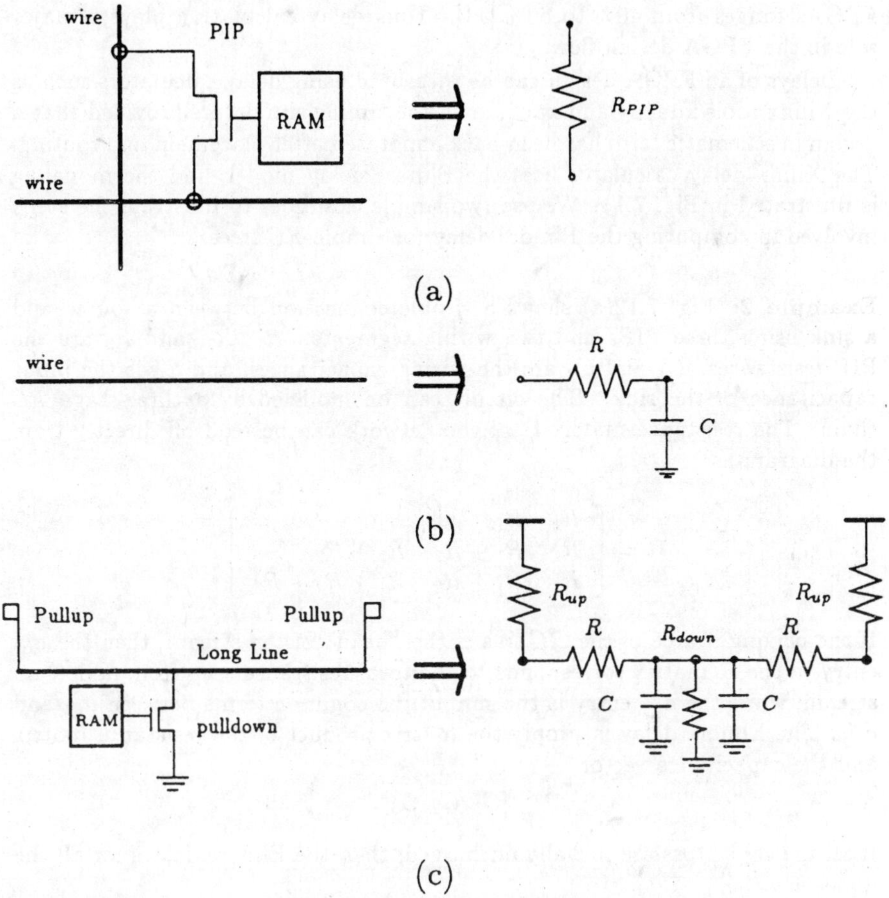

Figure 7.11: Resistor/Capacitor Delay Models for FPGAs: (a) an ON Pass Transistor is Modeled as a Resistor (b) a Wire Segment is Modeled as a Lumped Resistor Capacitor Network (c) a Long Line with Pullups is Modeled as a Lumped Resistor Capacitor Network.

7.6.1 Computing Signal Delay in RC Tree Networks

Unlike transistor-transistor logic, mask-programmed gate arrays, or custom VLSI chips, it has been estimated that the fraction of delay due to routing in FPGAs ranges from 40% to 60% [11]. Thus delay calculation plays a major role in the FPGA design flow.

Delays of an FPGA design can be measured using delay calculators such as the Xilinx tools **xdelay** and **xact**, or in the timing simulators provided that a design in schematic form has been back annotated after placement and routing. The Xilinx delay calculator uses the Elmore delay model, and the modeling is illustrated in Fig. 7.11. We use two simple examples to illustrate the steps involved in computing the Elmore delay for simple RC trees.

Example 2: Fig. 7.12(a) shows a simple connection between a source and a sink using three PIPs and two wiring segments. R_1, R_2, and R_3 are the PIP resistances. C_1 and C_2 are the wiring capacitances, and C_3 is the input capacitance of the sink. The circuit can be modeled by a three-stage RC chain. The resistance matrix $\mathbf{R}$ of the network can be read off directly from the diagram as

$$\mathbf{R} = \begin{bmatrix} R_1 & R_1 & R_1 \\ R_1 & R_1 + R_2 & R_1 + R_2 \\ R_1 & R_1 + R_2 & R_1 + R_2 + R_3 \end{bmatrix}$$

If the driving source of the RC tree is the "head" of the stream, then the $a_{i,i}$ entry in the $\mathbf{R}$ matrix corresponds to the total resistance seen from node i up stream, while the $a_{i,j}$ entry is the sum of the common terms between $a_{i,i}$ and $a_{j,j}$. The Elmore delay is simply the matrix product of the resistance matrix and the capacitance vector

$$[C_1, C_2, C_3]$$

If all the capacitors are initially uncharged, then the Elmore delays for all the nodes are

$$\begin{aligned} t_{d1} &= R_1(C_1 + C_2 + C_3) \\ t_{d2} &= t_{d1} + R_2(C_2 + C_3) \\ t_{d3} &= t_{d2} + R_3C_3 \end{aligned} \tag{7.3}$$

Example 3: Fig. 7.12(b) shows a three-node RC tree. The resistance matrix $\mathbf{R}$ of the network can be read off directly from the diagram as

$$\mathbf{R} = \begin{bmatrix} R_1 & R_1 & R_1 \\ R_1 & R_1 + R_2 & R_1 \\ R_1 & R_1 & R_1 + R_3 \end{bmatrix}$$

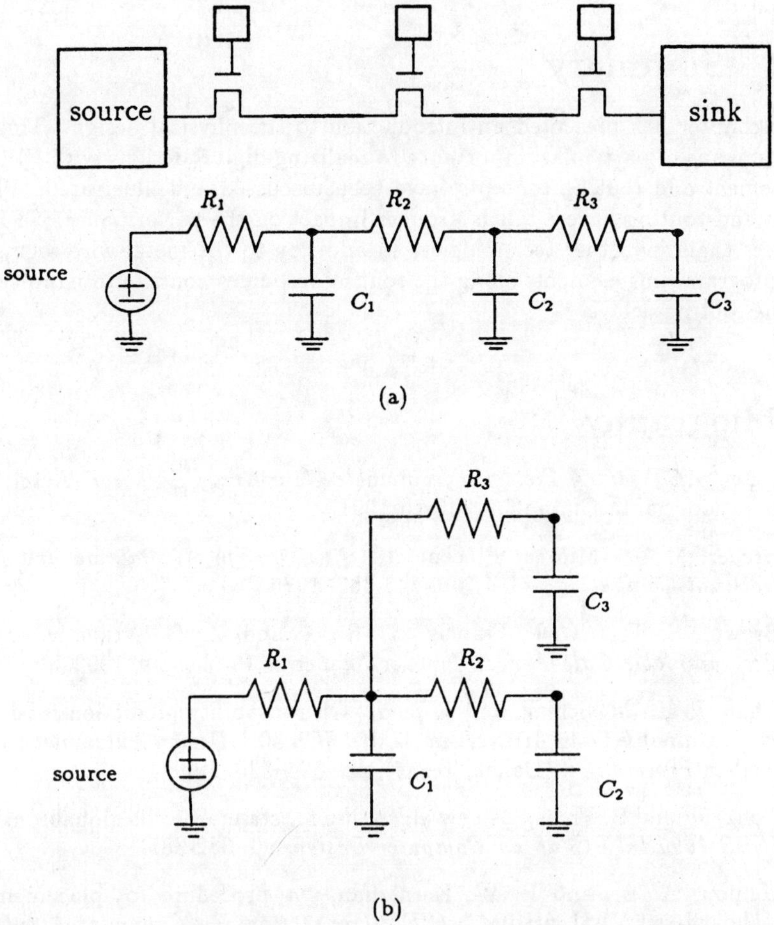

Figure 7.12: Examples of RC Trees.

Again, if all the capacitors are initially uncharged, then the Elmore delays for all the nodes are

$$\begin{aligned} t_{d1} &= R_1(C_1 + C_2 + C_3) \\ t_{d2} &= t_{d1} + R_2C_2 \\ t_{d3} &= t_{d1} + R_3C_3 \end{aligned} \tag{7.4}$$

7.7 Summary

This chapter has presented an introduction to the physical design. This design stage is of particular importance to realizing digital designs with FPGAs. Placement and routing concepts have been discussed and illustrated. Placement and routing have a much stronger impact on the performance of FPGA devices than any other technologies. In addition to the metal wire segments, the programming elements along the routing resources contribute significantly to the routing delays.

Bibliography

[1] Actel, *ACT family Field Programmable Gate Array Databook*, Actel Corporation, Santa Clara, CA, March 1991.

[2] Breuer, M.A., "Min-cut Placement," *J. of Design Automation and Fault Tolerant Computing*, vol. 1, pp. 343-362, 1976.

[3] Brown, S. D., R. J. Francis, J. Rose, and Z. G. Vranesic, *Field-Programmable Gate Arrays.* Kluwer Academic Publishers, 1992.

[4] Chan, P. K., M. Schlag, and J. Zien, "On routability prediction for Field-Programmable Gate Arrays," in *ACM IEEE* 30th *Design Automation Conference Proceedings*, (Dallas, Texas), pp. 326–330, June 1993.

[5] Cong, J., and B. Preas, "A new algorithm for standard cells global routing" *IEEE 1990 Int'l Conf. on Computer Design*, pp. 382-385.

[6] Dunlop, A. E., and B. W. Kernighan, "A procedure for placement of standard-cell VLSI circuits," *IEEE Transactions on Computer-Aided Design of Integrated Circuits and Systems*, vol. CAD-4, pp. 92–98, January 1985.

[7] El Gamal, A., "Two-Dimensional Stochastic Model for Interconnections in Master Slice Integrated Circuits," *IEEE Transactions on Circuits and Systems*, vol. 28, pp. 127–138, Feb. 1981.

[8] El Gamal, A., J. Greene, and V. Roychowdhury, "Segmented Channel Routing Is Nearly as Efficient as Channel Routing," *Proc. of the Advanced Research in VLSI*, UC Santa Cruz Conference, pp. 192-211, April 1991.

[9] Elmore, W. C., "The Transient Response of Damped Linear Networks with Particular Regard to Wideband Amplifiers," *Journal of Applied Physics*, vol. 19, pp. 55–63, Jan. 1948.

[10] Fiduccia, C. M., and R. Mattheyses, "A linear-time heuristic for improving network partitions," in *ACM IEEE* 19th *Design Automation Conference Proceedings*, (Las Vegas, Nevada), pp. 175–181, June 1982.

[11] Frankle, Jon, "Iterative and Adaptive Slack Allocation for Performance-Driven Layout and FPGA Routing" *Proc. of 29th ACM/IEEE Design Automation Conference*, pp. 536-541, June 1992.

[12] Hashimoto, A., and J. Stevens, "Wire Routing by optimizing channel assignment within large apertures," Proc. of the 8th Design Automation Workshop, pp. 155-163, 1971.

[13] Hollis, E. E., **Design of VLSI GATE Array ICs**, Prentice-Hall, Inc., 1987.

[14] Kernighan, B., and S. Lin, "An efficient heuristic procedure for partitioning graphs," *Bell System Technical Journal*, vol. 49, pp. 291-307, 1970.

[15] Kirkpatrick, S., C.D. Gelatt, and M.P. Vecchi, "Optimization by Simulated Annealing," *Science*, vol. 220, no. 4598, pp. 671-680, May 13, 1983.

[16] Lee, C. Y., "An algorithm for path connections and its applications," *IRE Trans. on Elect. Computers*, Vol. EC 10, Sept. 1961.

[17] Lengauer, Thomas, "*Combinatorial Algorithms for Integrated Circuit Layout*," John Wiley and Sons, 1990.

[18] Nilsson, N.J., *Problem Solving Methods in Artificial Intelligence*, McGraw-Hill, 1971.

[19] Palczewski, Mikael. "Plane Parallel A* Maze Router and its Application to FPGAs" *Proc. of the 29th Design Automation Conference*, pp. 691-697, June 1992.

[20] Preas, B.T., and M.J. Lorenzetti, *Physical design automation of VLSI systems*, The Benjamin Cummings Publisher, CA 1988.

[21] Rose, J., and S. Brown, "The effect of switch box flexibility on routability of FPGAs," *Proc. of IEEE 1990 Custom Integrated Circuits Conference*, paper No. 27.5, Sept. 1990.

[22] Sechen, Carl, *VLSI Placement and Global Routing Using Simulated Annealing*, Kluwer Academic Press, 1988.

[23] Small, B., A Bodmer, and S. Mourad, "Evaluating FPGAs: An experimental study," *Proc. of IPCCC*, pp. 134-140, 1993.

[24] Trimberger, S., and M.-R. Chene, "Placement-based partitioning for lookup-table-based Field-Programmable Gate Arrays," in *Proceedings of 1st International ACM/SIGDA Workshop on Field-Programmable Gate Arrays*, (Berkeley, California, USA), pp. 137–142, Feb. 1992.

[25] Xilinx, *The Programmable Gate Array Data Book*, Xilinx Corporation, San Jose, CA, 1992.

Chapter 8

Verification and Testing

8.1 Introduction

A key requirement for obtaining reliable electronic systems is the ability to determine that the systems are error-free [4]. This is known as *digital testing.* A circuit must be tested at several stages of its manufacturing and use to guarantee that it is working according to specifications. Such testing detects failures due to manufacturing defects and due to incorrectness in the design that escaped detection in the design validation phase. Test pattern generation is a complex problem [12]. Often, simulation patterns developed for design verification are augmented with patterns that are generated manually or by automatic test pattern generator (ATPG) to obtain a complete test set. A complete test set is capable of detecting all faults in the circuit. This test also verifies the functionality of its logic [1]. The patterns are then applied to the circuit using automatic test equipment (ATE). Because of the complexity of testing processes, a design approach aimed at making logic circuit more amenable to testing has been formulated. This approach to design is known as design for test (DFT) [20].

Testing a circuit prior to its implementation is known as *design verification.* At present, simulation is the most powerful tool for hardware verification. However, *formal verification* will, it is hoped, be available in the next generation of CAD tools.

In this chapter, we first distinguish between testing and verification from viewpoints of the manufacturers and the users of field-programmable gate arrays (FPGAs). Next we shall discuss the steps followed in validating designs implemented with FPGAs. The concepts of testing are presented in Section 8.4. The approach to testing FPGAs devices is the topic of Section 8.5. The remainder of the chapter will focus on design for test constructs used with finite state machines and printed circuit boards.

8.2 Verification Versus Testing

Designing with FPGA chips alleviates the testing process for the users. The unprogrammed FPGA devices are usually thoroughly tested at the factory. This is the task of the manufacturer or the vendor. The users, on the other hand, test the programmed device, which is actually design verification and validation.

Since FPGAs do not include any particular logic, the approach to their testing is not performed in the traditional way in which integrated circuits' testing is done. That is, the main problems in testing these devices are to verify that their programming elements are functional and the interconnect gaps between the routing segment are actually intact. The strategy in testing varies from one type of FPGA to another, depending on the architecture and the programming method. Thorough testing of the uncommitted device is not necessarily easy to accomplish, but once it is properly done, it is the same for all devices of the same type.

Most FPGA devices are implemented in the complementary metal-oxide semiconductor (CMOS) technology. They are electrostatic-sensitive devices. From the manufacturer to the programming site, there is a chance that the devices are damaged or degraded, for example, due to electrostatic discharge. Thus the users might check that the device is in working condition prior to actually programming it. After programming, the device, which is realizing the design, needs to be verified again. Here the patterns developed during simulation can be used for verification. Extensive simulation should be done particularly in the case of one-time-programmable devices such as the Actel and QuickLogic families. From manufacturing to programming, FPGAs pass through three testing (verification) stages: by the manufacturer, preprogramming (before implementation), and postprogramming (after implementation). Denoting the yield at the three stages as Y_m, Y_b, and Y_a, and assuming that the whole lot tested by the manufacturer is used by the same customer, the effective yield is then $Y = Y_m Y_b Y_a$. This yield can be estimated if testing data are available. It is equivalent to manufacturing yield for traditional ICs. In case of reprogrammable devices, $Y_a = 1$, since the design can be corrected and reimplemented.

Design validation can be done via simulation or formal verification. Both techniques make use of a certain model of the design under consideration. Unlike simulation, formal verification is based on proving correctness using formal logic. Although extensive research has been done in formal verification to date, there are little practical means to prove correctness of designs [22]. That is, the research has not yet materialized in a practical tool. Exhaustive simulation is equivalent to formal verification. However, it is not feasible for present complex circuits. Implementations in traditional ICs dictate that the user undertake thorough validation, simulation, and testing.

8.3 Verification

From the FPGA users' viewpoint, verification of the design is more relevant than is testing the unprogrammed device itself. This is particularly true for reprogrammable FPGAs. Verification can be done at different stages of the design cycle.

Consider the hierarchy of design flow discussed earlier in Chapter 3 and illustrated in Fig. 8.1 where different approaches to design entry are suggested. The design is verified by several simulations at different levels of its representation.

Whether the design has been entered as a schematic or as an hardware description language (HDL) model, it can be simulated with the appropriate high-level simulator. Some synthesis tools also support design verification at the gate level. After placement and routing, a design rule checker verifies that the design is indeed adhering to the logic block specifications and can actually be programmed. Then timing information, as characterized by the manufacturers, can be extracted after placement and routing of the design. The timing information is then *back annotated* to the original specification, either in schematic form or high-level language form, for simulation with timing information. A configuration format can be generated from the placed and routed design, which can be used to program the FPGA device. After programming the device, a user can even "read back" the configuration format as well as the internal states of an FPGA device for additional verification and validation.

8.3.1 Logic Simulation

Two types of simulations are used to verify the design: *functional simulation* and *timing simulation*. In functional simulation, the designer may verify the correctness of the design but not at operational speed. That is, no delays or at most nominal delays of the functional units are included. This is usually known as zero (or unit) delay. The main concerns are

- To check if each block performs its intended function.

- To modify the design to evaluate alternatives before finalizing the design.

The simulator translates the schematic into a netlist, which is a data structure listing all the gates and their connectivities. The designer prepares a set of inputs, verification patterns, and applies them to the circuit. The output is then examined by the user for correctness. Then if the designer is satisfied with the outcome, the translation should be performed. It is a good practice to use the same set of test vectors on the actual device to confirm that the circuit is a correct implementation of the design entered in the simulator.

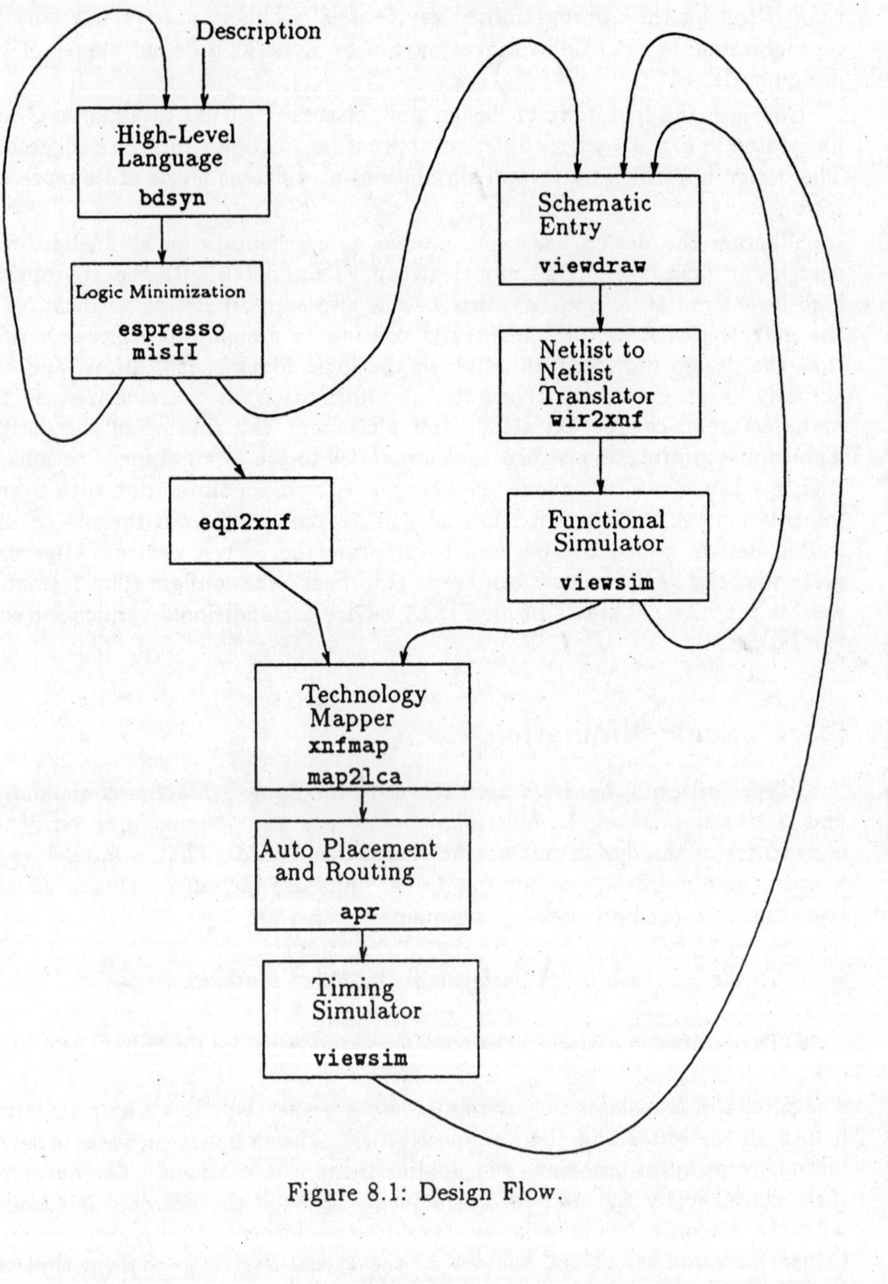

Figure 8.1: Design Flow.

8.3.2 Design Validation

The FPGA development system includes validation steps that are performed *before* and *after* the design's netlist is translated to the device configuration format. One would like to think the vendor's design implementation tools faithfully construct the design as specified by the users. However, there are some pitfalls. First, the design implementation tools can detect and report errors and warnings such as:

- Design exceeds the chip size.
- A net has no driver.
- Fanin or fanout is excessive.

In general, attention should be given to the error list; otherwise, the design will not be realized. Warnings should also be analyzed carefully since sometimes they may have repercussions in subsequent implementation steps. For example, in some cases, excessive fanin may be just a warning. However, it will eventually affect the circuit performance. Some design implementation tools attempt to "optimize" the design by removing all subcircuits that seem to be redundant, for example, logic modules that have unconnected outputs, two consecutive inverters, and so forth. A report on these actions is given and the designer needs to examine carefully before placement and routing. The major problem with this "optimization" is that the discarded subcircuit might affect the functionality or the performance of the design. It is usually not prudent to accept the modifications. Luckily, the designer has the option to turn off this optimization stage before it is performed.

Some FPGA development systems have verification tools for validation of the FPGA device configuration during and after configuration, for example, using cyclic redundancy codes (CRC) in the configuration format. After programming the device, a user can even "read back" the configuration format as well as the internal states of an FPGA device for further verification and validation.

8.3.3 Timing Verification

Timing verification can be performed in conjunction with functional verification or independently. The delays associated with the different gates are assigned. Usually nominal delays are assigned to the different gates (or functional units in the case of an HDL model). The delays are part of the library used in the design, or they may be modified. In this type of simulation, the net delays are not included, and thus the timing verification is not complete.

It is only when the design is placed and routed that the actual delays can be assigned to the gates and to their interconnects. Also, loading can be evaluated and assigned to the output pads.

Examples of simulation and back annotation are shown in Fig. 8.2. The design simulated is an 4-bit binary counter, `bcount4`. The figure shows the waveforms for the inputs, CLR (clear) and CLK (clock), and the outputs of all four positive edge-triggered flip-flops, Q(0) to Q(3). The first simulation was performed for functional verification with zero delays. The second simulation is performed on the circuit with the same input waveforms for CLR and CLK. It reflects the delays from clock to output, (CLK to Q). This delay is about one time unit.

bcount4 - logic simulation

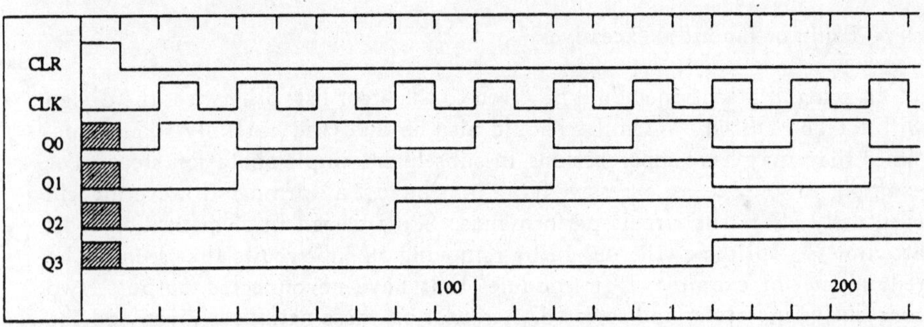

bcount4 - timing simulation

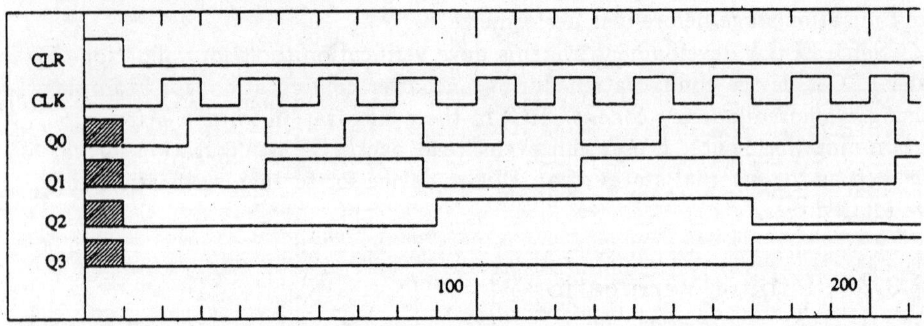

Figure 8.2: Logic and Timing Simulation.

8.4 Testing Concepts

8.4.1 Failures, Mechanisms, and Faults

Failures in integrated circuits can be characterized according to their duration, *permanent* or *temporary*, and by their mode, *parameter degradation*, or *incorrect design*. Figure 8.3 shows a classification of failures in digital ICs. *Permanent failures*, also called hard failures, are usually caused by breaks due to mechanical rupture or some wear-out phenomenon. They occur less frequently than *temporary failures*, which are failures that cannot be replicated [17]. These temporary failures, also called soft failures, are categorized as transient or intermittent. Transient failures are induced by some external perturbation such as power supply fluctuations or radiation. *Intermittent failures* are usually due to some degradation of the component parameters.

Fault Model	Description
Single stuck-at faults	One line takes the value 0 or 1.
Stuck-open faults	A failure in a pull-up or pull-down transistor in a CMOS logic device causes it to behave like a memory element.
Stuck-on faults	A transistor is always conducting.
Delay fault	A fault is caused by delays in one or more paths in the circuit.

Table 8.1: Most Commonly Used Fault Models.

Physical defects are due to different *failure mechanisms* that are largely dependent on the technology and even the layout of the circuit. Most of the manufacturing failure mechanisms are manifested on the circuit level as *failure modes*. The most common failure modes are open and short interconnections or parameter degradation. The different failure modes cause incorrect signal values. A *fault model* is the representation of the effect of a failure by means of the change that is produced in the system signal. Table 8.1 lists the common fault models. Fault models have the advantage of being a more tractable representation than physical failure modes, but they risk the omission of vital effects on system operations. For example, the most common fault model assumes *single stuck-at faults* even though it is clear that this model does not accurately represent all actual physical failures. The rationale for continuing to use a stuck-at fault model is the fact that it has been satisfactory in the past. However, the model is no longer sufficient for present circuits and technolo-

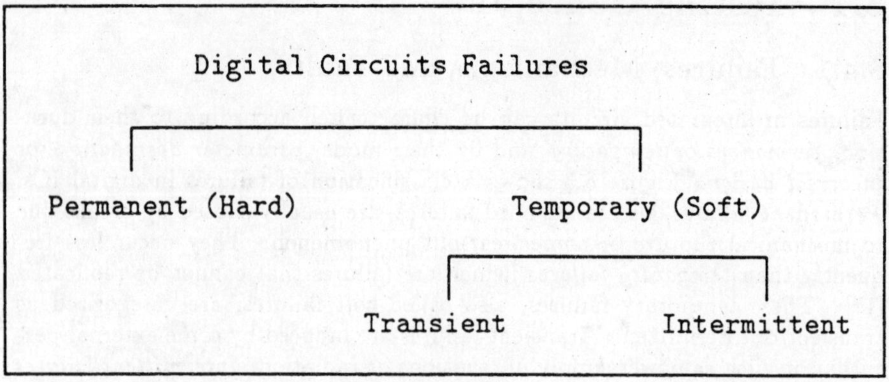

Figure 8.3: Classification of Digital Circuits Failures.

gies. With the advent of MOS technology, it has become evident that other fault models are needed to represent more accurately the failure modes in this technology [5].

8.4.2 Fault Coverage

The major goal of testing is to detect and locate faults in the circuit. Detection of faults involves two major concepts: *controllability* and *observability*. That is, the circuit has to be placed in a known state (controllability), and the effect of the fault has to be propagated to a primary output (observability). The appropriate combination on the primary inputs that allows the controllability and observability of the fault is called a *test pattern*. A set of test patterns is a *test set*. The number of patterns in a test set is know as the *test length*. Test pattern generation (TPG) is the process of developing test vectors to detect all detectable faults in a circuit. The effectiveness of the test sets is usually measured by the *fault coverage*. This is the percentage of the detectable faults in the circuit under test (CUT) that are detected by the test set. The set is complete if its fault coverage is 100%. This level of fault coverage is desirable but rarely attainable in most practical circuits. Moreover, 100% fault coverage does not guarantee that the circuit is fault free. The test checks only for failures that can be represented as a stuck-at fault model or the appropriate fault model used in the TPG process. Other failures are not necessarily detected. The fault coverage is calculated using a fault simulator. This is a logic simulator in which faults are injected at the appropriate nets of the circuits, usually one at a time. The response of the circuits to test pattern applications is compared with the good response of the circuit. The fault is considered detected if at least one of the patterns has a different response from the good circuit response.

8.4.3 ATPG Methods

The testing concepts discussed so far are illustrated with an example. The 2-to-1 multiplexer shown in Fig. 8.4 has three primary inputs, A, B, and S, and one primary output, F. The circuit has seven nets: A, B, S, S_1, G, H, and F. A *net* is a collection of gate input and output leads that are electrically connected. The net S consists of the primary input and two branches that fanin to the two AND gates. The stem and its two branches are considered three separate lines. That is, one of the branches might be faulty without affecting the stem or the other branch. However, any fault on the stem will definitely affect the branches. Also, faults on the branch entering the inverter are equivalent to those on S_1. According to this view of net decomposition, we shall consider only faults on the lines A, B, S_1, S_2, G, H, and F. Each line may be stuck-at zero (SA0) or stuck-at one (SA1). A total of 14 stuck-at faults is possible.

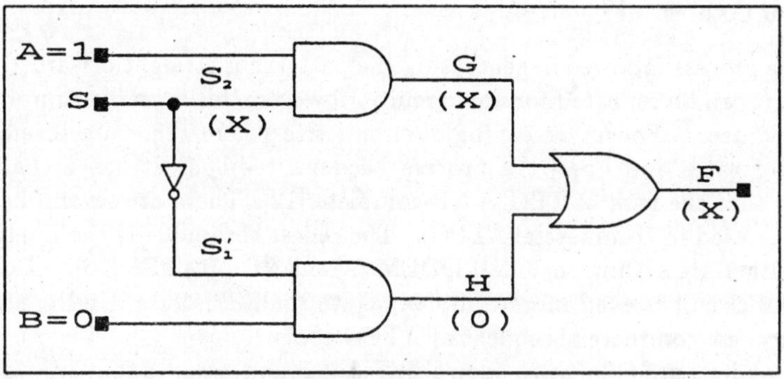

Figure 8.4: Stuck-at Faults in a 2-to-1 MUX.

Assume that branch S_2 entering the upper AND gate is faulty. Let the faulty signal be indicated by X, where $X = 0$ if the node stuck at 0 or $X = 1$ if it is stuck at 1. To sensitize the fault to the primary output F, we need (1) to place logic 1 on the signal A (the fault then appears on G) or (2) to hold the other input to the OR gate, H, low by applying logic 0 to B. Thus the value of $AB = (10)$ enables the observation of a fault on S_2. To determine if this node is actually SA1 (or SA0), we need to control it to zero (or to one). In summary, the pattern that detects S_2 SA1 is $ABS = 100$ and that which detects the SA0 is $ABS= 101$. In the same manner, we can generate patterns for all the other faults. Table 8.2 lists the patterns and the faults they detect where $S/1$ means line S stuck at 1.

The 2-to-1 MUX has a 4-pattern-long test. Notice that

- One pattern may detect several faults that form an equivalence class.

Pattern ABS	Detected Faults
100	$S_2/1$, $B/1$, $H/1$, $G/1$, $F/1$
101	$S_2/0$, $G/0$, $A/0$, $F/0$
010	$B/0$, $H/0$, $S_1/0$
011	$A/1$, $S_1/1$, $G/1$, $H/1$, $F/1$

Table 8.2: A Minimal Test Set for a 2-to-1 MUX.

This test serves in detecting that the circuit is faulty but does not give any diagnosis as to which fault caused the failure.

- Some faults can be detected by more than one pattern, for example, $F/1$ or G/1.

The process followed in generating the patterns is straightforward. Conceptually, it can be repeated for any circuit. However, this is an oversimplification of the process. For any meaningful circuit size where there are several thousands of gates and lines, the process becomes tedious. It has actually been proven that the task of TPG is NP-complete [12]. There are several heuristics that are used in commercial ATPGs. The oldest technique is the D-algorithm [15]. Other algorithms include PODEM [9] and SOCRATES [16]. TPG for sequential circuit is even more complex, as we shall illustrate it next, and there are very few commercial sequential ATPG tools.

Consider an SR latch as shown in Fig. 8.5 and assume that we would like to detect the stuck-at fault $A/0$. To sensitize the fault to the primary output Q, we must have $R = 0$. To control the line A, we assert $Q = S = 0$. It is easy to place the required values on S and R since they are primary inputs. But to force $Q = 0$, first make $R = 1$. Thus to detect the fault $A/0$, it is necessary first to apply the pattern $(SR) = (01)$ and then follow by the pattern $SR = (00)$. The patterns have to be in this proper sequence. A sequential circuit needs first to be placed in a certain state before sensitizing (propagating) the fault to the output. In most cases more than two patterns are needed. For example, if we need to detect a SA0 fault on the ripple carry out of a 32-bit counter, it is necessary to apply 2^{31} clock cycles.

8.4.4 Types of Tests

Since a test pattern is a combination of primary inputs, it is conceivable to use all possible combinations (an exhaustive test set) and apply them to the circuit. This exhaustive approach to testing has the advantage of being easy to

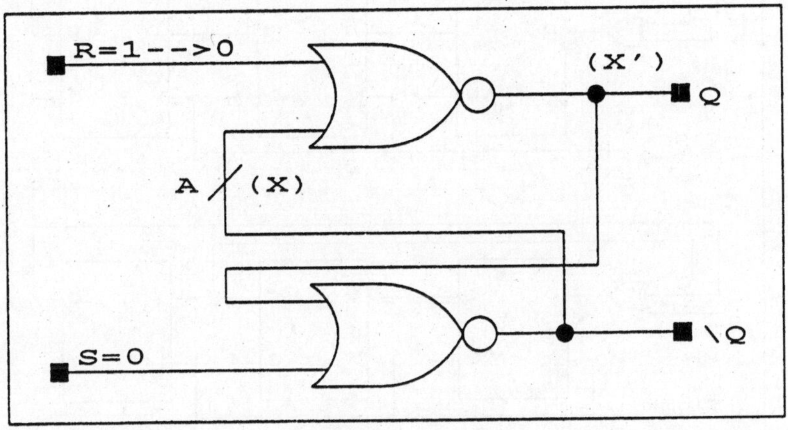

Figure 8.5: An *SR* Latch.

generate and of yielding 100% fault coverage. However, such testing method is efficient only for purely combinational small circuits (no more than 20 primary inputs) [13]. It should be clear from the previous section why exhaustive testing is not applicable for sequential circuits. For sequential circuits, the sequence in which patterns are applied is crucial for fault detection. The test patterns have to be applied in the required sequence to guarantee fault detection. In circuits where each primary output is a function of only a subset of primary inputs, it is possible to apply an exhaustive test on this subset of inputs. This is known as *verification testing* [14]. Another alternative is to segment the circuit in subcircuits and to test the segments exhaustively.

Test patterns may also be generated at random. The cost of generating the test is minimal, but a fault simulator is needed to *grade* the test and assess the fault coverage. The advantage of random testing [3] is that it has been shown to detect up to 80% of the faults. Thus many commercial ATPGs use random testing as a first stage of the test pattern generation and apply heuristics to deal with the still undetected faults. In purely random testing, a test pattern may be generated more than once. However, pseudorandom generation is more appropriate to ensure that there is no repetition of patterns. Pseudorandom test sets may be generated by a software program or by a linear feedback shift register (LFSR) in hardware. Fig. 8.6(a) shows a three-stage LFSR that can generate all 3-bit combinations but the all-zeros pattern. The number of feedbacks and their positions determine the length of the test.

Testing of the hardware is usually implemented by instruments that are called automatic test equipment [8, 18]. This is a very important phase of

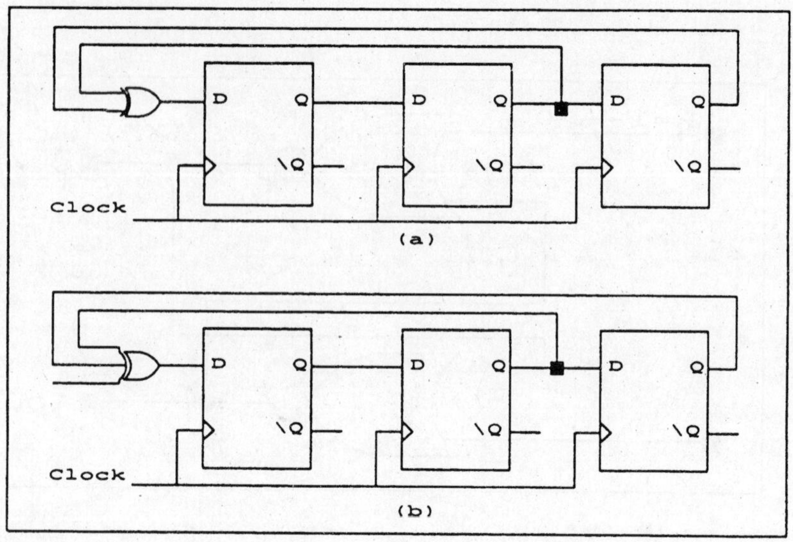

Figure 8.6: An LFSR (a) as an ATPG, (b) as a Signature Analyzer.

digital testing, but it will not be discussed in this book.

8.5 Testing FPGAs

Designing with FPGAs alleviates the testing process for the users. The unprogrammed FPGA devices are usually thoroughly tested by the manufacturers [21, 2]. Since FPGAs are field programmable by definition, their testing involves examining their logic blocks, programming elements, and interconnects. FPGAs have special circuitry to facilitate their testing. For example, the ACT devices have · *obes*, and the Xilinx devices have *read backs*.

In the RAM-based Xilinx chips, the testing strategy includes

1. Writing and reading back all configuration cells
2. Pseudoexhaustive testing of all CLBs
3. Continuity and short test on the interconnect metal segments [21]

The different testing configurations exercise the device without removing it from the tester.

Testing the device in this fashion is equivalent to programming it for all possible configurations. Users can apply some of these testing methods before programming the device for their application, for example, connecting all the flip-flops into one or more shift register configuration and checking that they can shift known patterns. A similar approach is followed in the case of other reprogrammable devices using EPROM and EEPROM technologies. Testing all configurations is problematic for EPROM since erasing the device is not on the fly as in the case of EEPROM- and RAM-based FPGAs.

Testing a one-time-programmable device is likely to be more difficult. Actel devices include extra hardware dedicated for testing that does not require programming the antifuses except those dedicated for testing. But this feature is not accessible to the designer. As in the case of erasable floating gate–based devices, there is really no guarantee that all potential interconnect sites are actually connected. The only sure way to find out is to test the device after programming. Such problems will be uncovered as testing for *programmability failures* is performed. The Actel devices have a probe that allows the observation of all nets after programming [7]. Testing a programmed ACT1 device requires only test patterns to control the nets.

8.6 Programmability Failures

In addition to failures common to conventional ICs, FPGAs are subject to unique failure modes that we identify as *programmability failures*. These types of failures can be caused by a faulty chip or the malfunctioning of the device programmer. Like any electronic device, device programmers can fail. For example, a noncalibrated device programmer may apply too little or too much voltage for incorrect periods and cause an Actel's antifuse to remain intact or to connect with a higher resistance than specified [10]. This might cause missing and extra connections between the segments of the routing wires. In addition, they may introduce longer delays in the connecting wires.

Each FPGA is prone to specific physical defects depending on the underlying architecture and programming technology. For instance, in EPROM-based technology, erasure of programmed connections can occur with exposure to light. The logic cell array's use of volatile static memory makes it prone to problems caused by noise, radiation, and power dropouts. Independently of the technology, failure modes in the interconnect between metal wire segments may result in extra connections or missing connections. Depending on the position of this extra or missing connection, the failure results in different possible fault models. Examples of mapping these types of failures into known fault models are listed in Table 8.3.

Failures due to a programmer's errors include using the incorrect programming data file or using the wrong blank FPGA type. These types of failures can result in logic block functional failures as well as interconnect failures.

Connection Failure	Fault Model
• Extra connections between independent wires	Bridging fault
• Missing connection on the lead that is an input to a gate	Stuck-open fault
• A bridging fault where one of the wires is VCC or GND	Stuck-at fault
• Higher resistance of a connection [10]or too long or too short a connection	Delay fault

Table 8.3: Mapping Failure Modes into Fault Models.

8.7 Design for Testability

Design for test constructs are usually incorporated in digital circuits to increase the controllability and observability of the different nodes of the circuits [20]. The controllability is the degree of ease or difficulty of placing the circuit in a known state. The observability is the ability of observing the different nodes of the circuit at the primary outputs. There are some "commonsense" approaches to DFT that are usually referred to as ad hoc techniques:

- Partition the design into blocks that can be tested independently.
- Add test point that can increase the observability.
- Provide an external clear to all flip-flops.
- Design a large counter out of smaller modules or make it loadable.

Here we shall present three main DFT constructs: scan path for sequential circuits, boundary scan that facilitates board testing, and built-in self-test (BIST). These constructs can be implemented in FPGAs at manufacturing time or at customization time by the users. For FPGAs, we distinguish between two approaches to DFT: *preapplication* DFT and customized DFT. Preapplication DFT facilitate testing in the architecture of the unprogrammed FPGA or for any implemented design. As was mentioned in the previous section, every type of FPGA has special circuitry to facilitate testing. Not all the structures are made public. DFT that can be used in conjunction with the customized chip is available only in Xilinx 4000 series. The XC4000 FPGAs have a boundary scan architecture that facilitates board-level testing. It is also possible to add BIST features.

8.7.1 Scan Path Design

In scan path design all the flip-flops of the circuit can be connected in the form of one or several shift registers [19]. The contents of the shift register(s) are shifted in and out to facilitate the controllability and observability of the circuit. The reconfiguration of the flip-flops in a shift register is accomplished by inserting a 2-to-1 MUX at the input of each flip-flop. This is illustrated in Fig. 8.7, where the box labeled "combinational" represents all the combinational components of the circuits and SCAN-IN and SCAN are two extra inputs. If the control SCAN is low, the circuit operates in normal mode. But, when SCAN is high, the flip-flops are connected in a shift register. In the shift mode, the first flip-flop receives its signal directly from the input, SCAN-IN. In this manner, we can easily place the circuit in any desired state (controllability). Then, the internal state of the circuit is observed by shifting out the contents of the register (observability). Thus the inputs and the outputs of the flip-flops are considered, respectively, primary outputs and primary inputs of the circuit.

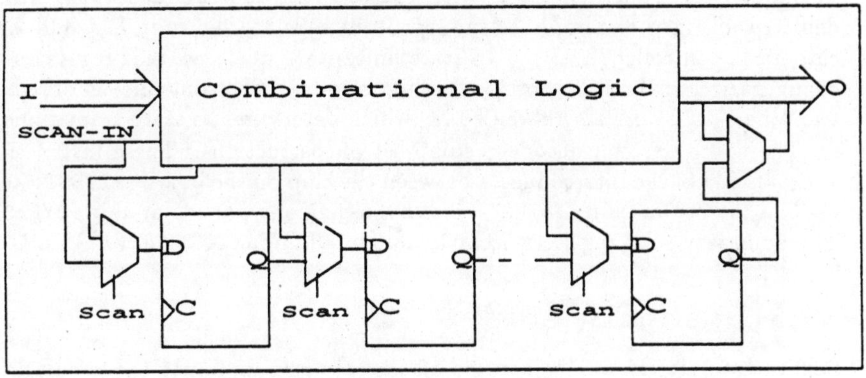

Figure 8.7: Scan Path Design.

Scan path testing involves

1. Testing the flip-flops.
2. Testing the combinational part of the circuit, which consists of

   (a) SCAN=0, set initial state into the shift register.

   (b) SCAN=1, apply a test pattern and clock the circuit once.

   (c) SCAN=0, clock the circuit N time, where N is the number of flip-flops in the chain to shift the register out, while shifting in the register the data for the next state through SCAN-IN,

   (d) return to step b.

This DFT structure has effectively partitioned the combinational component of the circuit to subcircuits. The outputs of these subcircuits are the inputs to the flip-flops or the primary outputs. Thus the complexity of test pattern generation has been greatly reduced. The costs of this reduction are the possible delays incurred by the 2-to-1 multiplexers and the increase in test data and test application time on the ATE equipment.

A special case of this DFT technique is known as *level sensitive scan design* (LSSD) [6]. In such a case the design is implemented with double-latch instead of flip-flops.

8.7.2 Boundary Scan Design

The most difficult aspect of board-level testing is determining whether or not the ICs pins on a printed circuit board are properly connected. For this, a scheme that permits connecting all I/O of all chips on the board in a shift register was devised [11]. To allow the formation of this register, all the primary inputs and outputs enter the circuit of a chip via flip-flops that can be strung in chain to form a shift register with `TDI` as input and `TDO` as output. This is depicted in Fig. 8.8. Simplified input/output cells are shown in Fig. 8.9. Each chip has a controller (TAP), a 16-state finite state machine, that regulates the configuration of the flip-flops in the shift register. The controller is driven by two other global signals `TMS` and `TCK`, which determine its state transitions.

The controller can decode a small set of instructions. These instructions allow testing of the interconnects between the chip on the board (`EXTEST`), testing of a particular chip (`INTEST`), and bypassing any particular chip (`BYPASS`). The boundary scan registers are available in all Xilinx XC4000 FPGAs.

8.7.3 Built-in Self-test

So far, test patterns were generated off the chip and applied to the circuit using ATE equipment. As circuits grow more complex, they become more difficult to test and BIST techniques become more attractive. In built-in testing, the test generation and the response verification are both done on the chip or the board. In this case random testing is used. Fig. 8.6(a) shows a random pattern generator. It is a finite state machine consisting of N flip-flops, few XOR gates (MOD 2 addition), and feedbacks to the individual flip-flops. The LFSR cycle is determined by its *characteristic polynomial.* This polynomial is defined by the positions of the feedbacks. At most the cycle consists of $2^N - 1$ states. Usually, pseudorandom test sets are too long. To compact the circuit's response to a long test set, an LFSR is also used as a *signature analyzer*. A signature analyzer is shown in Fig. 8.6(b). As the test sequence is applied, the response passes through the LFSR. At the end of the application, the internal state of the LFSR represents the *signature* of the circuit under the test applied. The main disadvantage of signature analysis is *aliasing*. Aliasing occurs when the signature of the circuit for a given fault is equal to the signature of the

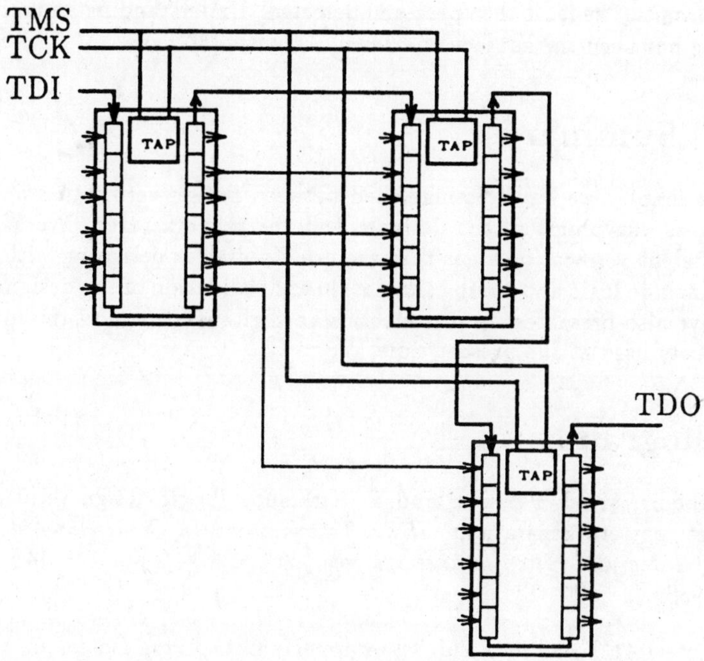

Figure 8.8: Boundary Scan Testing.

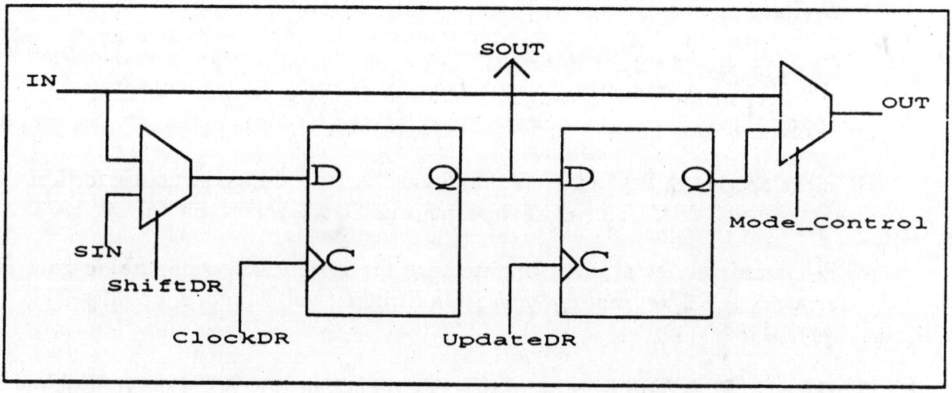

Figure 8.9: An I/O Port in Boundary Scan Design.

good circuit. The fault then passes undetected. Understanding and minimizing aliasing has been the subject of extensive studies [3].

8.8 Summary

In this chapter we have distinguished between design verification and digital testing as viewpoints of the designer and the manufacturer. We have listed the different verification steps that are usually used in designing with FPGAs. They include logic and timing simulation and validation of the design realized. We have also presented an introduction to digital testing and the main DFT constructs used at the present time.

Bibliography

[1] Abadir, M., J. Furgson, and T. Kirkland. "Logic design verification via test pattern generation," *IEEE Transactions on Computer-Aided Design of Integrated Circuits and Systems, Vol. CAD-7,* pp. 138–149, January 1988.

[2] Actel. *ACT Family Field Programmable Gate Array Databook.* Actel Corporation, Santa Clara, CA, 1991.

[3] Bardell, P. H., W. H. McAnney, and J. Savir. *Built-in Pseudorandom Testing of Digital Circuits.* New York: John Wiley, 1987.

[4] Breuer, M. A., and A. D. Friedman. *Diagnostics and Reliable Design of Digital Systems.* Computer Science Press, 1976.

[5] Case, G. R. "Analysis of actual fault mechanisms in CMOS logic gate." *In Proc. of the 13th Design Automation Conference*, pp. 265–270, June 1976.

[6] Eichelberger, E. B., and T. W. Williams. "A logic design structure for LSI testability." *IBM Journal of Research and Development*, pp. 90–99, 1977.

[7] El-Gamal, A., et al. "An architecture for electrically configurable gate arrays," *IEEE Journal of Solid-State Circuits*, Vol. 24, no. 2, pp. 394–398, April 1989.

[8] Feugate, R. J., and S. M. McIntire. *Introduction to VLSI Testing.* Englewood Cliffs, NJ: Prentice Hall, 1988.

[9] Goel, P. "An implicit enumeration algorithm to generate tests for combinational logic circuits," *IEEE Transactions on Computers*, Vol. C-30, no. 3, pp. 215–220, 1981.

[10] Hamdy, et al. "Dielectric based antifuse for logic and memory IC." *Paper presented at the Int'l Electron Device Meeting*, pp. 786–788, May 1988.

[11] Maunder, C., and F. Beenker. "Boundary scan framework for structured design for test." *In the Proc. Int'l Test Conf.*, pp. 714–723, 1987.

[12] Miczo, A. *Logic Testing and Simulation*. New York: Wiley, 1986.

[13] McCluskey, E. J. "Verification testing — A pseudoexhaustive test technique," *IEEE Transactions on Computers*, Vol. C 33, no. 6, June 1984.

[14] McCluskey, E. J., and S. Bozorgui-Nesbat. "Design for autonomous test," *IEEE Transactions on Computers*, Vol. C 30, no. 11, pp. 866–875, 1984.

[15] Roth, J. P. "Diagnostics of automata failures: A calculus and a method," *IBM Journal of Research and Development*, pp. 278–281, 1966.

[16] Schultz, M. H., et al. "Socrates: A highly efficient automatic test pattern generation system," *IEEE Transactions on Computer-Aided Design of Integrated Circuits and Systems, Vol. CAD-7*, no. 1, pp. 126–137, 1988.

[17] Siewiorek, D. P., et al. "A case study of C.mmp, Cm* and C.vmp." *In Proc. of the IEEE*, Vol. 66, pp. 1178–1220, October 1978.

[18] Stevens, A. K. *Component Testing*. Reading, MA: Addison-Wesley, 1986.

[19] Williams, M. J. Y., and J. B. Angel. "Enhancing testability of large scale integrated circuits via test points and additional logic," *IEEE Transactions on Computers*, Vol. 1, C-22, pp. 46–60, 1973.

[20] Williams, T. W., and K. Parker. "Design for testability — a survey," *IEEE Transactions on Computers*, Vol. 1, C-31, pp. 2–15, 1982.

[21] Xilinx. *User Guide and Tutorials*. Xilinx Corporation, San Jose, CA, 1992.

[22] M. Yoeli, ed. *Formal Verification Hardware Design*. Parsippany, NJ: IEEE Society Press, 1990.

Chapter 9

Design Guidelines and Case Studies

9.1 Introduction

In this chapter, we shall use several simple design examples to illustrate the design guidelines, design principles, pitfalls, and considerations behind using the reprogrammable field-programmable gate array (FPGA) technology to implement digital systems.

Some important practicalities and details will be considered first; then one small design example is presented followed by a medium-scale and interesting design.

9.1.1 Design Guidelines

We advise you to keep the architecture of the FPGA constantly in mind. For instance, when you are doing an FPGA-based design, remember to

1. Check the number of pins in and out of a logic cell (block) and the fanout of nets.
2. Check the width of look-up table of a RAM-based FPGA, the number of inputs of a multiplexor in the MUX-based FPGA. It is because the performance of a technology mapper is sensitive to these parameters. Don't expect the design tools to be "smart" enough to take care of all the problems for you.
3. Use "guide flags" in the schematic drawing to tip the technology mapper (e.g., CLB, CLBMAP, FMAP, GMAP, HMAP) on how to group the logic together into cells. Please use guide flags such as "critical" with great caution; don't overdo it. It doesn't help to flag everything critical. Keep in mind that the "critical" flag of a net is only a tip to inform the router to route the net first, among many nets.

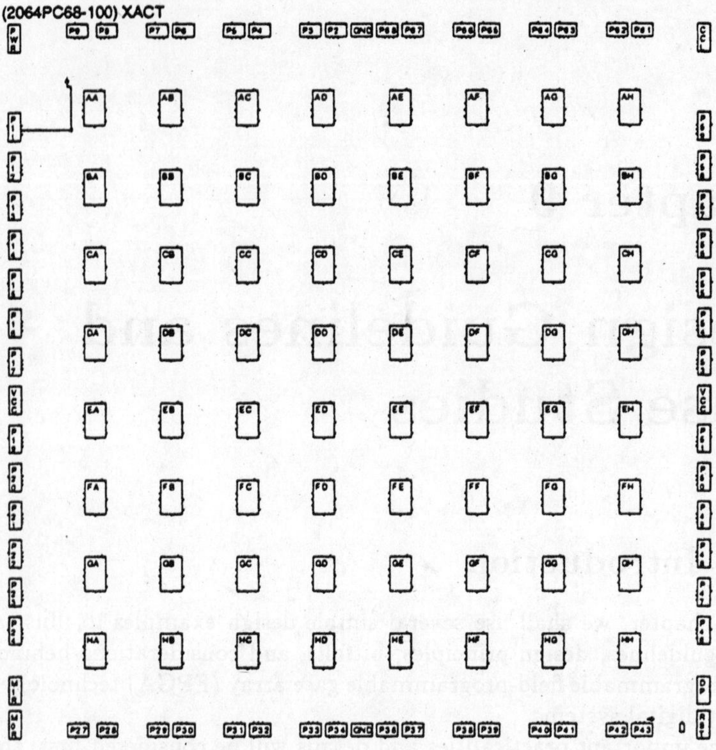

Figure 9.1: Global Clock Buffer in a Xilinx FPGA.

4. Count the number of high-fanout nets and make sure that this number is less than what is available to support high-fanout nets on the device.
5. Check the average pins-to-logic-cell ratio, as some routing-limited FPGAs cannot complete the placement and routing phase if the pins-to-logic-cell ratio is too high [3].
6. Avoid locking down the I/O pins. Let the placement and routing tool chooses the pin assignment for you, at least for the first placement and routing round.

There are pros and cons to the practice of exploiting the architecture of the FPGA devices. Your design will not be very portable, or will be less retargetable to a different technology. It is because the library cells of your retarget technology are very likely to be different from the FPGA cell library. The other side of the coin is the fact that your FPGA design uses fewer cells to implement and may run faster. You can't win them all, it is your choice!

9.1.2 Download, Clocks, Global Buffers, and Internal Oscillator

The reprogrammable XILINX FPGAs generally can be configured in several ways. For the purpose of prototyping, it is wise to configure your FPGA parts by downloading the bitstreams directly from a computer. You can use either the download cable or the `xchecker` [1]. The older download cables uses a parallel port and transmits the configuration bits *serially*. The download cable has no error detection nor correction, so the bitstream is suspectable to corruption by noise, in which case may fail to configure the FPGAs. It is advised to avoid hanging the download cable over components that can interfere the cable (e.g., a clock signal) during download.

It is vital to use the on-chip global clock buffer to distribute the clock edge uniformly with low skew to all the flip-flops inside the logic blocks. In a XC3000 FPGA, the global clock line goes only to the "K" pin of the Configurable Logic Blocks (CLBs). Normally you don't have to worry about this if you enter a design through a schematic capture program using the Xilinx library. However, it is a different story if you are using a lower-level entry tool such as the `xact` design editor, for performance reasons, to layout your design. You need to plan ahead and allocate the routing resources carefully to maximize the performance of your design. Figure 9.1 shows the location of a global buffer (GCLK) which is at the upper left-hand corner of an XC2000 FPGA by Xilinx. There are four global primary and four secondary buffers in the XC4000 series.

There is a second global buffer at the lower-right-hand corner of all Xilinx XC3000 FPGAs, which is called ACLK. It can be used in conjunction with the internal CMOS oscillator, an external crystal, two capacitors, and a resistor to generate a clock signal up to 20 MHz. In the Xilinx XC4000 FPGAs, an internal multivibrator generates signals of frequency range from 15 Hz to 8 MHz. These signals can be used as time bases. But keep in mind that they will drift with change of temperature and vary with process variations.

9.1.3 Tristate Busses, Horizontal Long Lines, and Vertical Long Lines

Figure 9.2 shows the locations of the tristate buffers relative to a CLB in a Xilinx XC3000 series FPGA. Each horizontal routing channel has two rows of tristate buffers spanning the columns, with the exception of the horizontal channels on the north and south ends. The outputs of the tristate buffers that belong to the same row feed a horizontal long line. These tristate buffers which is available at the schematic drawing level as `TBUF`, can be used to implement shared busses or the wired-AND (WAND) logic function.

Strong pullups and weak pullups are available for the tristate bus. The pullups are located at the far ends of the horizontal long lines.

Other than for implementing the functions mentioned above, long lines can also be used to route high fanout signals. Typically, the placement and routing

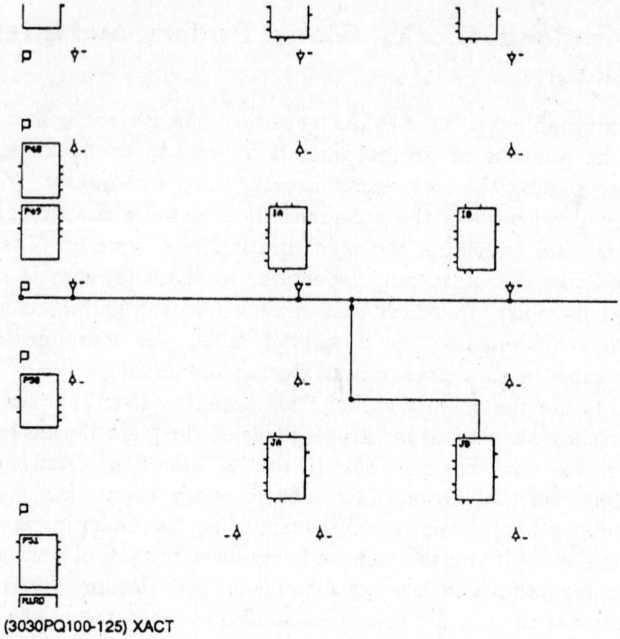

Figure 9.2: Tristate Buffers and Horizontal Lines in a Xilinx 3000 series FPGA.

tool starts using long lines for routing if the fanout of a net exceeds certain value, for example, 7. It is interesting to note that the long lines can actually be split (in all XC3000 series devices except the XC3020) into pieces to allow them to carry more nets.

9.1.4 Pad Assignment, Conflicts, and Programming Pins

Field-Programmable Gate Arrays have both dedicated and shared pins to program or configure the devices. Dedicated programming pins are not the major concern except that they must be tied to the correct logic level. Shared pins have dual identity in the sense that they are used to indicate programming mode or status during configuration but will become a user pin afterward. To avoid conflicts, attention should be paid to avoid contention of pins that are used as inputs during configuration and become outputs in normal operation. For example, the `M2` pin of the XC3000 series LCA is of this nature. A 4.7K ohm pullup resistor must be used if the LCA device is used in the slave mode. This pin must never be connected to output pin of any other device that is active before the FPGA is configured. Another example is the `M2`, `M1` and `M0` pins of the XC4000 series LCA. These pins are used to control the different programming modes during the configuration of the LCA devices. After a de-

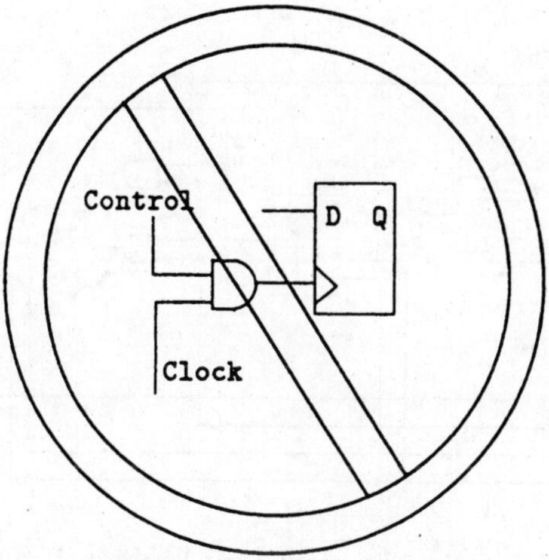

Figure 9.3: No Gated Clocks!

vice has been configured, these pins can be used for reading the internal states of the LCA device. Resistors combined to form voltage dividers can be used to ensure the proper states of these pins before and after device configurations.

9.1.5 Avoid Gated Clock

You should feed the clock line directly to *ALL* clock inputs of flip-flops. It is because gated clocks, as shown in Fig. 9.3 create timing problems by inserting delays and skews in the clock line. Consequently, the components in the system will be out of synchronization. Imagine that all the instrument players in an orchestra start to ignore the conductor and play in their own rhythm! Gated clocks also introduce the potential for race conditions and spurious pulses which cause false clocking. To retard the activity of a flip-flop according to a control signal, you should use the "flip-flop enable" to condition the flip-flop if you really need to. This flip-flop is sometimes referred to as the "E-type" flip-flop. It *holds* a data item until it is triggered by "enable" to clock in another item. One can implement this feature with two XC2000 logic blocks by using some inputs of the combinational logic block in conjunction with the internal feedback path of a CLB. This "enable" or hold feature is available directly in the XC3000 and XC4000 families as the EC input of a CLB, and appears as "CE" clock enable in some flip-flops and counters in the macro libraries.

Exercise 1: Map the "Left-Right Register with Hold" as shown in Fig. 9.4 onto XC2064 CLBs. Draw a diagram to illustrate the configuration of the CLBs and the connections between the CLBs. Do you lose any performance (in CLBs and delays) with the XC2064 as opposed to using XC3020? If yes,

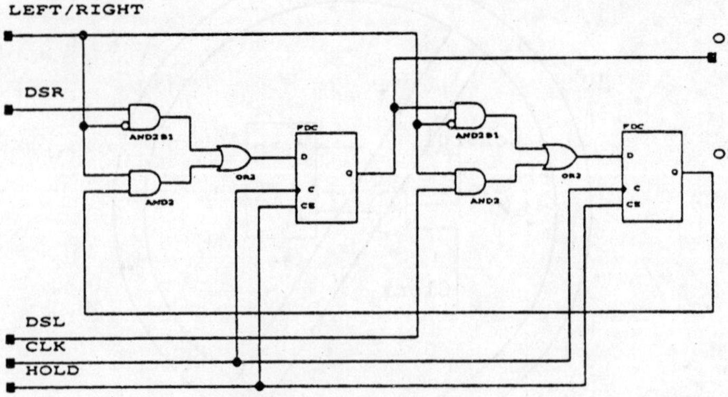

Figure 9.4: A Left-Right Shift Register with Hold.

by how much?

9.1.6 Avoid Asynchronous Designs

A very common novice mistake is to introduce asynchronous elements in the system, either knowingly and unknowingly. We shall illustrate what to avoid, the basic principle, and the proper method with the design of a binary counter.

Example 1: The Tale of Two 5-bit Counters.

Suppose we need a 5-bit counter as part of a digital system to be implemented using Xilinx FPGAs. We browse through the standard libraries/macros and the closest that we can find are 4-bit counters. The component `C16BCRD` is a 4-bit binary counter with reset. There are 3 inputs – the clock `C`, clock-enable `CE`, asynchronous reset `RD`, five outputs `Q0`, `Q1`, `Q2`, `Q3`, and the terminal count `TC`. It doesn't seem unreasonable to consider building a 5-bit counter with `C16BCRD` and an additional toggle flip-flop. Since the terminal count signal `TC` is asserted (HIGH) when the 4-bit counter `C16BCRD` reaches 15, one would be tempted to use this signal to clock the toggle flip-flop. Figure 9.5 shows such a bad design.

This circuit doesn't work because it violates the principle of single-phase synchronous design:

All the flip-flops should be connected to the same clock.

The problem of this design is illustrated in Fig. 9.6, which shows that the toggle flip-flop `Q4` is false triggered by the `TC` signal of the 4-bit binary counter when it reaches 12 (`Q3=1,Q2=1,Q1=0,Q0=0`). The source of this problem is, once again, due to a problem in differential in propagation delay as a result of

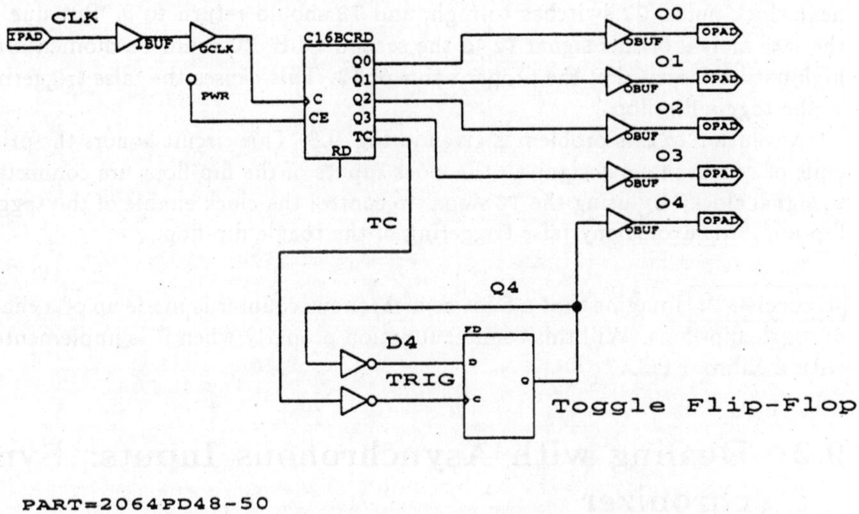

Figure 9.5: An Asynchronous Binary Counter.

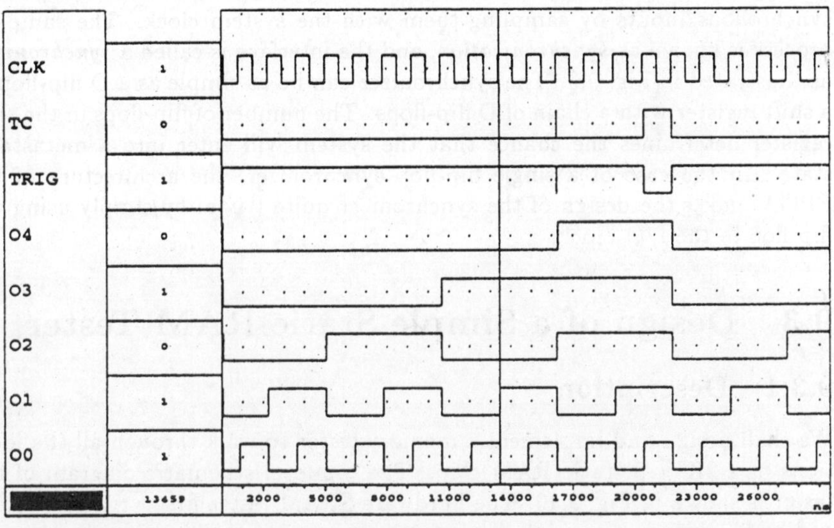

Figure 9.6: Timing of the 5-Bit Asynchronous Counter Showing the Triggering of the Toggle Flip-Flop Due to a Glitch of T2 at 16000ns.

technology mapping. We reveal the construction of the 4-bit binary counter C16BCRD in Fig. 9.7. Notice that the signal TC is the "and" of Q3, Q2, and T2. The signal T2 in turn is the "and" of Q1 and Q0. When the binary counter reaches eleven (Q3=1,Q2=0,Q1=1,Q0=1), T2 is asserted. Upon the arrival of the next clock pulse, Q2 switches to high, and T2 should return to 0. But due to the late arrival of this signal T2 to the second CLB, TC could be momentarily high until the arrival of the proper value of T2. This causes the false triggering of the toggle flip-flop.

A solution to this problem is given in Fig. 9.8. This circuit honors the principle of synchronous design: all the clock inputs of the flip-flops are connected to signal clock. By using the TC signal to control the clock enable of the toggle flip-flop, this avoids any false triggering of the toggle flip-flop.

Exercise 2: Imagine that a 5-bit asynchronous counter is made up of a chain of toggle flip-flops. Will this counter function properly when it is implemented with a Xilinx FPGA?

9.2 Dealing with Asynchronous Inputs: Synchronizer

A common approach to designing a sequential circuit with asynchronous inputs is to use a synchronous sequential circuit and place an interface between it and the asynchronous inputs. The interface changes the asynchronous inputs to synchronous inputs by sampling them with the system clock. The sampling process is known as *synchronization*, and the interface is called a *synchronizer*, as illustrated in Fig. 9.9. The synchronizer can be as simple as a D flip-flop or a shift register with a chain of D flip-flops. The number of flip-flops in the shift register determines the chance that the system will enter into a metastable state. In the case of a single flip-flop synchronizer, the architecture of the FPGA makes the design of the synchronizer quite trivial by simply using the flip-flop in the I/O block.

9.3 Design of a Simple Static RAM Tester

9.3.1 Description

We shall design and implement a memory tester to walk through all the locations of a 2K × 8 static RAM chip. The top-level schematic diagram of the tester is shown in Fig. 9.10. The hardware SRAM tester first writes a pattern and then immediately reads back from the same location. The tester then compares the two values. The patterns are derived from the lower 8 bits of an address counter that scans the memory chip. Some special patterns can be designed for detecting shorts between data lines and address lines. This will

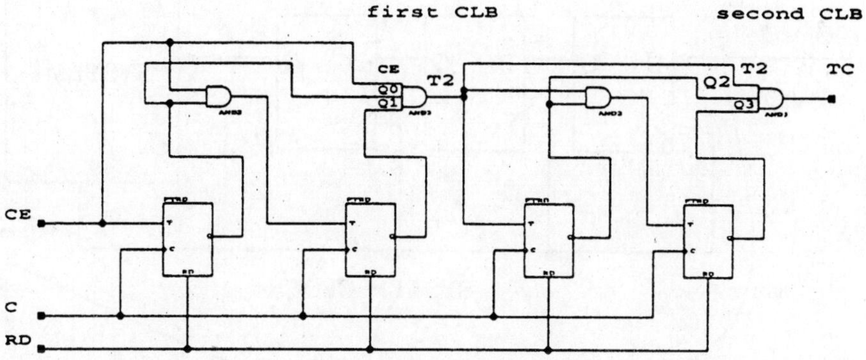

Figure 9.7: The Internals of the 4-Bit Counter C16BCRD.

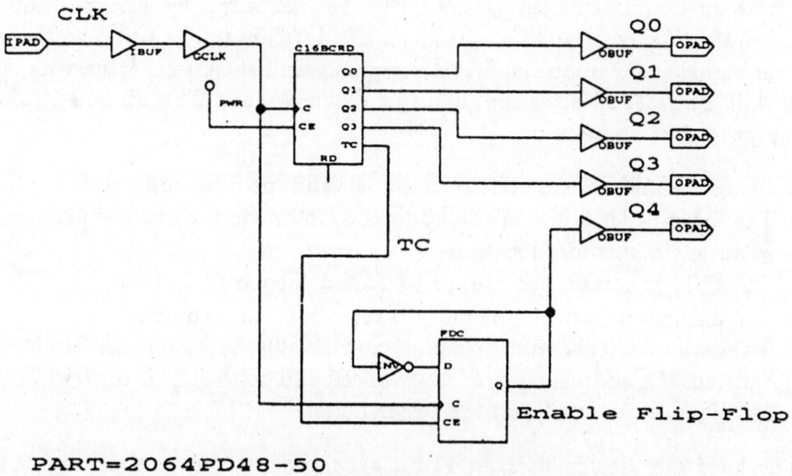

Figure 9.8: A Synchronous 5-Bit Binary Counter.

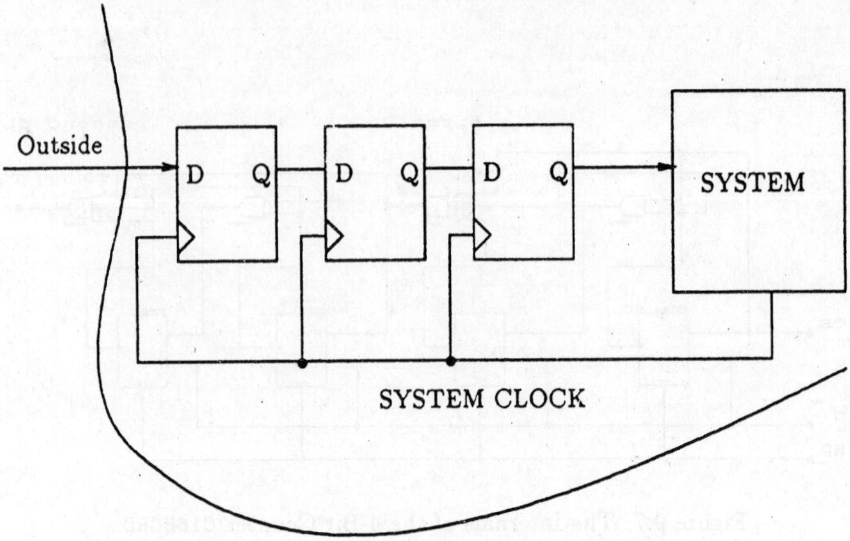

Figure 9.9: A Simple Three-Stage Synchronizer.

be left as an exercise for the reader. For the meantime, we assume that all faults are manifested as stuck-at faults on the data lines. The tester will assert an error signal (`ERR`) upon the first discrepancy and then halt. Otherwise, the tester will assert `PASS` after completing a flawless scan. This simple memory testing procedure consists of

1. Using the address counter to supply a valid memory address
2. The lower-order 8 bits of the address counter are used as the pattern to write to the memory location
3. The FSM generates the proper *write* and then *read* signals
4. The data read from the memory is forwarded to a comparator
5. If the two patterns match (equal) and the 2K address space is not exhausted, the address counter is advanced and go to step 1; otherwise, the `ERR` signal is asserted and the tester halts.

The hardware design uses an 11-bit synchronous counter (see Fig. 9.11) to generate the memory address. The counter is made up of three smaller synchronous counters from the macro library. The terminal count signal `TC` indicates that the 11-bit counter has overflowed. An 8-bit comparator is used for comparing the values written and read from the memory location (see Fig. 9.12).

A note on the write-cycle timing requirement using the *write-enable controlled* (as opposed to *chip-select controlled*) mode: `WR` is the last to be asserted and the first to be negated when enabling the RAM's internal write operation. The `CS` and `WR` signals theoretically can be simultaneously negated after being

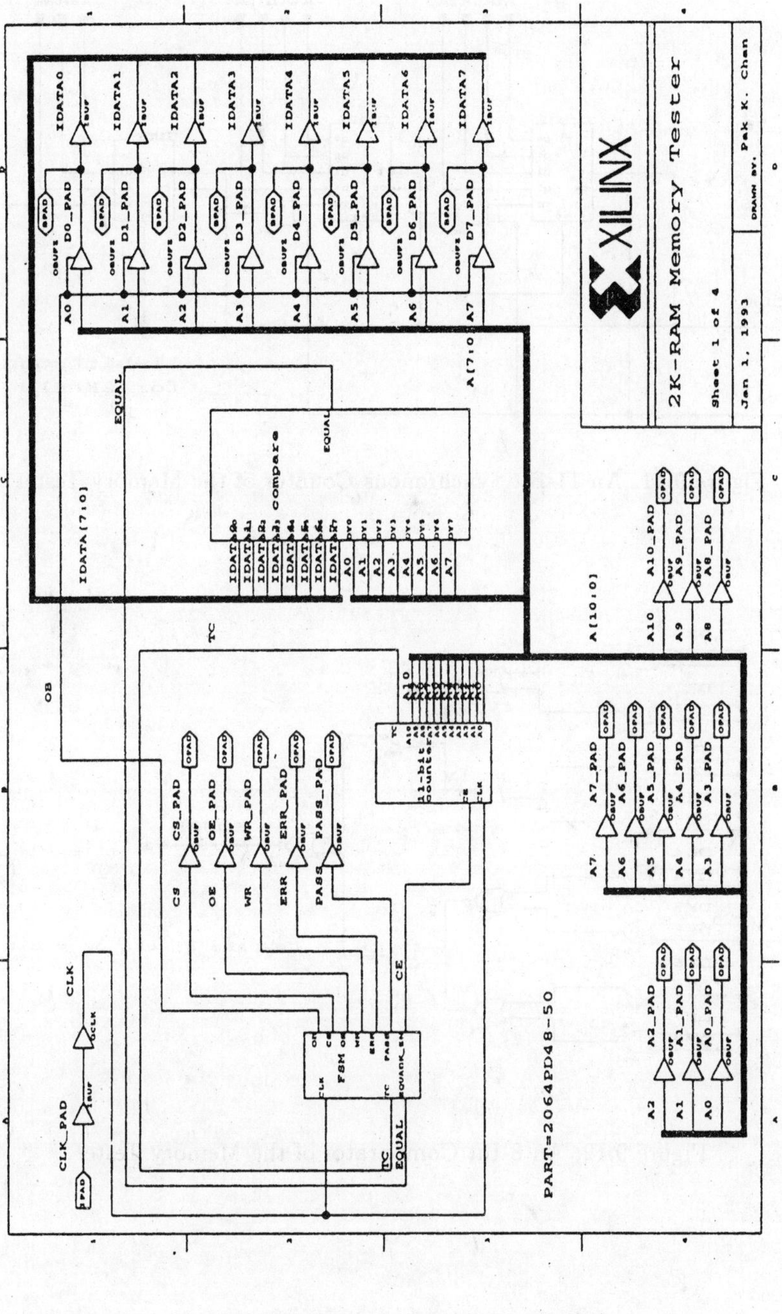

Figure 9.10: Top-Level Schematic of a 2K × 8 Memory Tester.

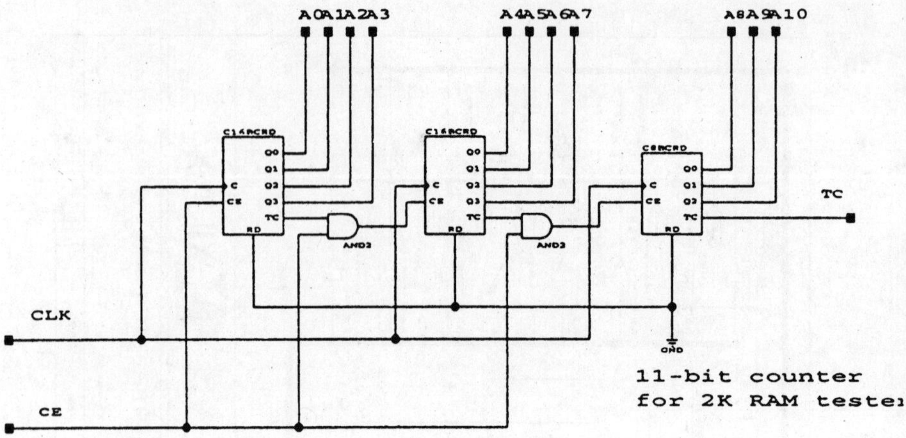

Figure 9.11: An 11-Bit Synchronous Counter of the Memory Tester.

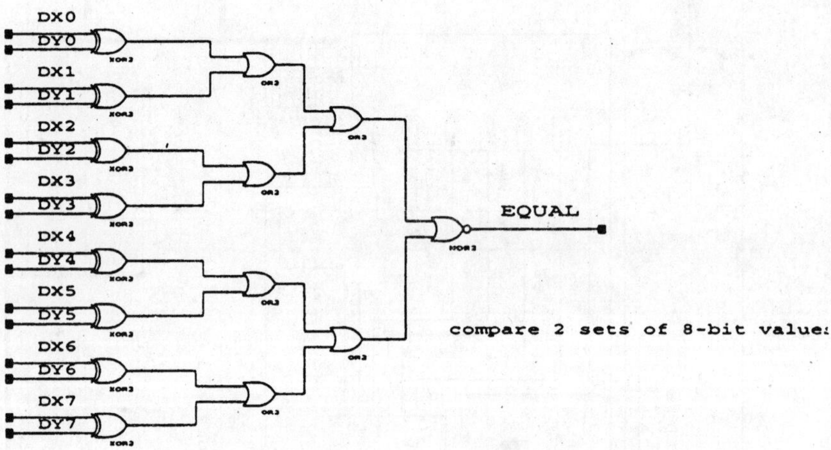

Figure 9.12: An 8-Bit Comparator of the Memory Tester.

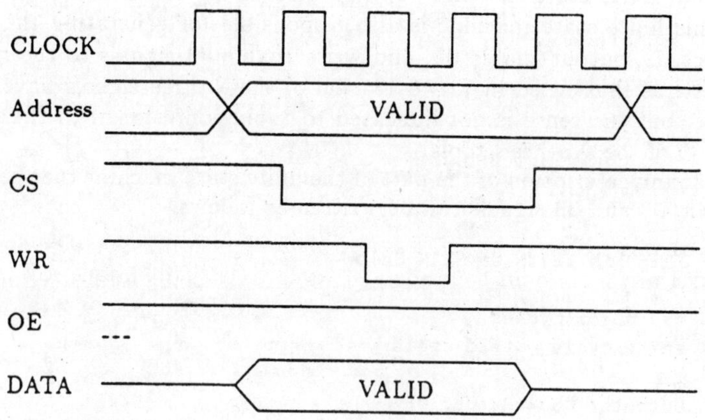

(a) WRITE TIMING

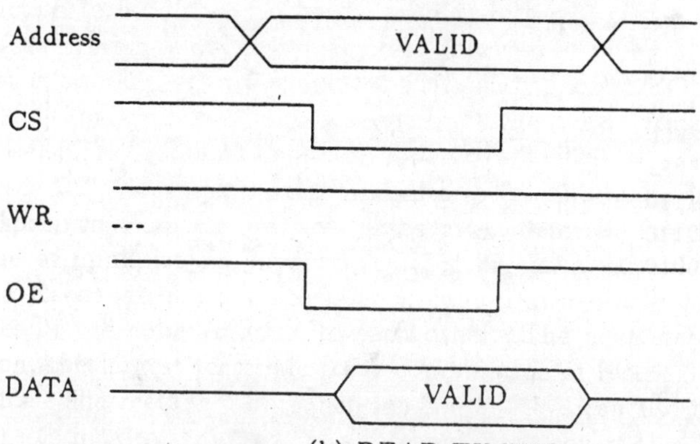

(b) READ TIMING

Figure 9.13: Read/Write Timing of the Memory Tester.

asserted. The FSM has been designed to ensure that data are strobed into the RAM by WR well before the RAM is disabled by the chip select CS, as shown in Fig. 9.13. Without this extra precaution, it is very possible that the RAM may be disabled before the arrival of the write strobe due to gate and routing delays. The clock in Fig. 9.13 is drawn for reference; the memory doesn't require a clock.

A small finite state machine (FSM) is needed to coordinate the memory scan. This finite state machine is also responsible for generating the proper chip-select CS, output-enable OE, and write enable WR signals to the memory under test, as illustrated in Fig. 9.13. All of these three signals have active-low sense, but the sense is not indicated to avoid confusion with the naming convention of the synthesis tools.

A `mustang` description of the part of the finite state machine that generates the CS, WR, OE and OB signals can be written as follows:

```
#   This FSM tests an SRAM chip.
#
#   two major cycles
#   write cycle   read cycle
#
#   Outputs: CS, WR, OE, OB
.i 1
.o 4
.s 8
#
# write cycle
#
#          CWOO
#          SREB
- S7 S6   1110
- S6 S5   0110
- S5 S4   0010
- S4 S3   0110
- S3 S2   1111
#
# read cycle
#
- S2 S1   0101
- S1 S0   0101
- S0 S7   1111
```

However, after encoding the states using the one-hot encoding scheme and minimizing the combinational logic of the state machine with `misII` using the standard script, the logic expressions correspond to these control signals are given in Equations (9.1).

$$
\begin{aligned}
INORDER &= PS7\ PS6\ PS5\ PS4\ PS3\ PS2\ PS1\ PS0; \\
OUTORDER &= NS6\ NS5\ NS4\ NS3\ NS2\ NS1\ NS0
\end{aligned}
$$

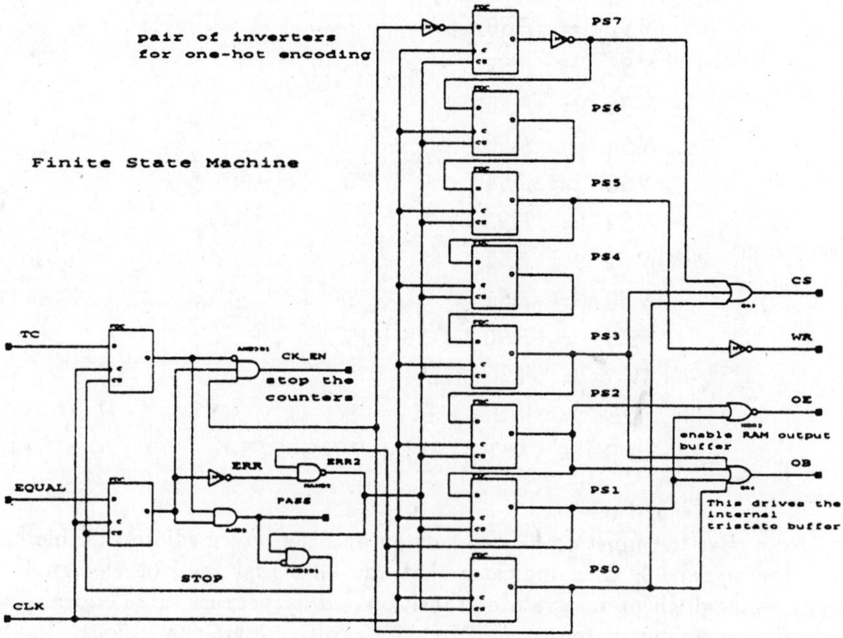

Figure 9.14: The Finite State Machine of the Memory Tester.

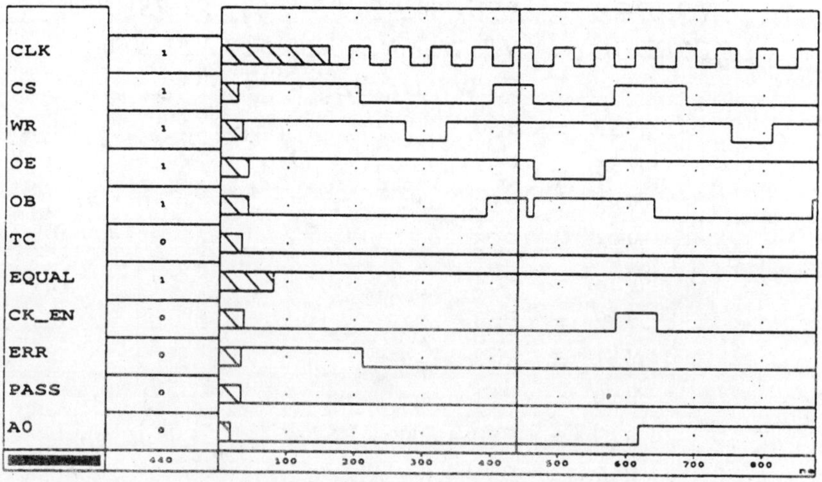

Figure 9.15: Simulation Results Showing the Memory Control Signals.

$$
\begin{aligned}
& CS\ WR\ OE\ OB; \\
NS7 &= PS0 \\
NS6 &= PS7 \\
NS5 &= PS6 \\
NS4 &= PS5 \\
NS3 &= PS4 \\
NS2 &= PS3 \\
NS1 &= PS2 \\
NS0 &= PS1 \\
CS &= PS0 + PS3 + PS7 \\
WR &= [32] + OB \\
OE &= [32] + PS5 \\
OB &= PS0 + PS1 + PS2 + PS3
\end{aligned} \tag{9.1}
$$

Note that the function `WR` depends on and the intermediate variable [32], and the signal `CS`. This indicates that the `WR` signal (and other signals as well) will "glitch" during state transition. It is because `misII` minimized this Boolean network represented in their positive logic expressions. In this application, it would be more appropriate to represent the signals in their complements since all of them are active-low. By inspection, we can rewrite the Boolean network as given in Equations (9.2).

$$
\begin{aligned}
INORDER &= PS7\ PS6\ PS5\ PS4\ PS3\ PS2\ PS1\ PS0; \\
OUTORDER &= NS6\ NS5\ NS4\ NS3\ NS2\ NS1\ NS0 \\
& \quad CS\ WR\ OE\ OB; \\
NS7 &= PS0 \\
NS6 &= PS7 \\
NS5 &= PS6 \\
NS4 &= PS5 \\
NS3 &= PS4 \\
NS2 &= PS3 \\
NS1 &= PS2 \\
NS0 &= PS1 \\
CS &= PS0 + PS3 + PS7 \\
WR &= PS5' \\
OE &= (PS1 + PS2)' \\
OB &= PS0 + PS1 + PS2 + PS3
\end{aligned} \tag{9.2}
$$

An implementation of the finite state machine of the memory tester based on these expressions is shown in Fig. 9.14. Timing simulation results showing the memory control signals is given in Fig. 9.15. Note that the signal `WR` has clean transitions since it depends on only one state variable $PS5$. This would not be possible using the logic expressions minimized by `misII` as shown in Equations (9.1).

The rest of the FSM is for halting the machine depending on the values of the terminal count `TC` from the counter or the `EQUAL` signal for the comparator. Unlike conventional logic design technique, we didn't use a "dead" state for halting. We exploit the clock-enable `CE` feature available in the FPGA, by asserting the clock-enable `CL_EN` low to halt the finite state machine. The complete memory tester design takes 30 XC2000 CLBs.

Exercise 3:

The memory tester shown in Fig. 9.10 assumes that all faults are manifested as stuck-at faults at the data lines. Modify the memory tester to facilitate the detection of single stuck-at faults at the address lines.

9.4 Design of a Tetris Machine

The memory tester is a small project that can be designed, implemented, and tested within a day or so. To demonstrate the ability of FPGAs to implement larger digital systems, we shall use a larger example: a Tetris machine. The following Tetris machine was designed by Martine Schlag.

9.4.1 Description

We shall design and implement a digital machine to play Tetris. The implementation will use one Xilinx `XC3020-PC84`, one Xilinx `XC3042-PC84`, and a 16-byte static RAM. The design will interface with a "host" computer that will be responsible for displaying the Tetris screen and keeping the machine's score. The Tetris machine will be solely responsible for keeping the status of the Tetris bucket; the machine cannot inquire about the status of the bucket from the host computer.

Tetris is a two-dimensional bin-packing game invented by Russian computer scientist Alexander Porkofiev. Figure 9.16 shows the Tetris "bucket." Seven different Tetris tiles, as shown in Fig. 9.17, are randomly chosen and presented one at a time. The player must decide where to place each tile in a rectangular 20 (rows) ×8 (columns) bucket to achieve the maximum packing of tiles. The player can rotate the tile counter-clockwise in units of 90 degrees and move the tile laterally left or right to the desired location. The tile then drops to the bottom of the bucket until it is stopped by tiles already in the bucket. Figure 9.16 shows the bucket after several tiles have been placed in the bucket. If the player succeeds in completely filling any row in the bucket, then that

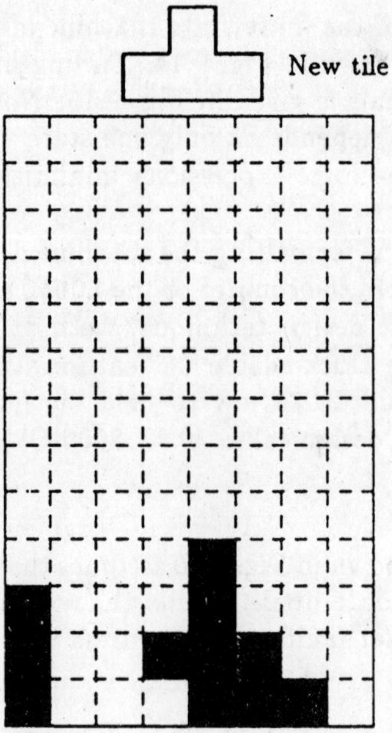

Figure 9.16: A Tetris Bucket (Showing Only 13 Rows; the Actual Game has 20 Rows).

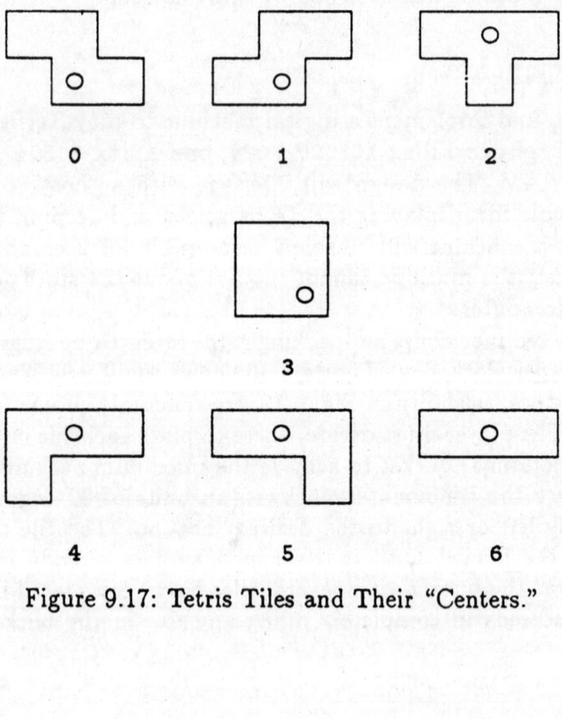

Figure 9.17: Tetris Tiles and Their "Centers."

row is deleted and all rows above it move down one row. The game continues until any column of the bucket becomes completely full and no more tiles can be placed. The height of tile "6" and the size of the Tetris bucket are the only small differences between our Tetris game and the Nintendo Tetris game.

9.4.2 The Game Environment

Our Tetris player machine will interface to a host PC that presents tiles one at a time, as illustrated in Fig. 9.18. We shall provide a 16-byte SRAM as a scratch pad for the Tetris machine. The SRAM will be used to store the contents of the Tetris bucket. The host PC runs a driver that informs the player of the next tile type, gives the player some "think time," processes the player's (machine) decision of where to place the tile, displays the bucket, and displays the game score. This driver is included in the appendix. The host PC presents a new tile (encoded in 3 bits as `newtile`) to the Tetris machine. The rotation of the tile, which is in increments of 90 degrees, is encoded in 2 bits as `rotation`. The size of the Tetris bucket is 20 rows tall by 8 columns wide. This allows the encoding of the lateral position of a tile in 3 bits (`lateral`), with the position measured with respect to the center of the tile. An active-low signal `Your_move` informs the Tetris machine that a new tile is presented. When the Tetris machine has its move ready, it provides the `rotation`, and `lateral` values and then asserts `move_ready` low.

9.4.3 Approach and Strategy

Strategic Planning

First, we devise and evaluate some basic strategies by a high-level simulation. To examine how successful our strategies are, we code the strategy(ies) in the `C` programming language and integrate into a `C` program. We record the scores achieved by each strategy. There is a pitfall that the high-level language constructs in `C` are too powerful and may not be implemented efficiently or directly with the given hardware constraint. So we must constantly keep the implementation constraints in mind - the design must be realizable in one XC3020 and one XC3042 Xilinx FPGAs.

Judge Feasibility

We judge the feasibility of a strategy by estimating of the number of CLBs and the number of flip-flops used to implement the strategy. We allow a 10% margin of error in our estimation. A naive way is to enter the whole design, and then place and route it. It is quite discouraging to find out that your design won't fit with the available FPGAs after going through this process. There are some rules of thumb to estimate the feasibility in terms of capacity requirement of your design. If your design can be represented in some intermediate form

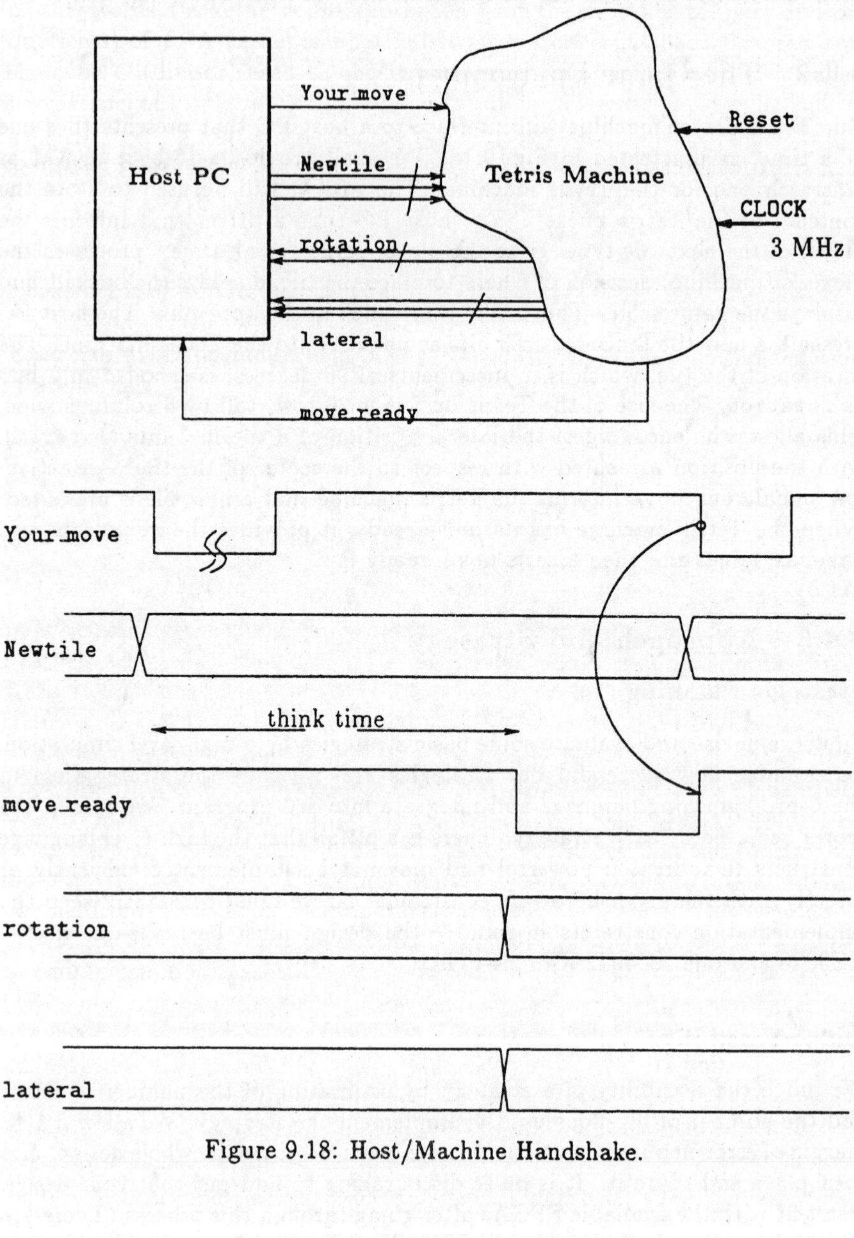

Figure 9.18: Host/Machine Handshake.

like the Boolean equation format, then we can use a rule of thumb mentioned in Chapter 5:

$$\text{Number of CLBs} \approx \frac{\text{Literal Count}}{5}$$

The next thing to check is the number of flip-flops required by your design with the number of flip-flops available on the XC3000 series devices. Also, for routability reasons, try not to use more than 85% of the total number of cells available.

Iterate

If there are unused CLBs, we can use them to improve the strategy, that is, go back to the strategic planning step. On the other hand, if a sophisticated strategy consumes too many CLBs, then we should try a simpler strategy.

9.4.4 Where to Land a Tile?

We shall briefly describe the operations of the Tetris machine in this and the next subsection. The logic design of the machine will be elaborated in Section 9.5. The XC3000 FPGA architecture does not provide any on-chip RAM, so an *image* of the status of the Tetris bucket is maintained in an external SRAM. However, we shall implement all the counters and registers that are associated with the updating of the status of the Tetris bucket inside the FPGAs.

The basic algorithm exercised by the Tetris machine is quite simple, as illustrated in Fig. 9.19. The first phase is to clear the first 16 bytes of the SRAM to represent an empty Tetris bucket, in which case all the counters and registers inside the FPGAs are also cleared. The second phase is to evaluate and determine the placement of a tile. Given a Tetris tile and a location to land the tile, the machine evaluates the *best* rotation of the tile. It tries all possible locations to land the tile, and it updates the Tetris bucket afterward. The steps in the second phase are repeated upon every new tile presented.

Since no Tetris tile spans more than three columns, we search for a perfect match between a tile and three adjacent columns by trying all four different rotations. The *best* location is where there is a perfect match, or the lowest point in the Tetris bucket in the absence of any match. Also, to avoid having the tiles building up on one particular side of the bucket, the search is alternating between scanning from left to right and vice versa. Now, to evaluate the best location to land a given tile, the Tetris machine uses eight height registers (`H[1],...,H[8]`) to reflect the current maximum heights of the eight columns of the bucket. For example, the Tetris bucket as depicted in Fig. 9.16 has heights of $\{3,0,0,2,4,2,1,0\}$ from left to right. Since the Tetris bucket is of size 20×8, we need eight 5-bit registers to record the exact heights of the eight columns. However, it is observed that it would be hard to revive the game once the heights of the columns went beyond 16. We balance the performance of the Tetris machine against the amount of hardware that is needed

```
Tetris machine
  Phase 1.Initialize the bottom 16 rows of the Tetris bucket
        Height_registers = Lrow = Crow = 0
  Phase 2.For each tile presented by the host:
              step A. determine new height for each column
                      and all rotations
              step B. retain the best column and rotation
              step C. update the height registers
              step D. update Tetris bucket in SRAM,
                      remove all filled rows
              step E. inform host the column and rotation desired
        endfor
```

Figure 9.19: Pseudo-Code for Tetris Machine Algorithm.

to implement a perfect machine. Accepting this trade-off, we use only 4-bit `height_registers` (instead of 5 bits).

9.4.5 Updating the Tetris Bucket

After the determination of where to place the tile, the next step is to update the image of the Tetris bucket in the RAM. Since filled rows have to be eliminated by the rules of the game, the image of the Tetris bucket is stored rowwise as 16 rows with 8 bits per row. Also, a tile is no larger than 3 by 3, so no more than 3 rows will be need to be updated at each round. The steps to update a row are first to generate a row mask, then read the row from the memory, and write the masked row back to the memory. Rows that are all "1's" are eliminated according to the rules of the game.

9.5 Logic Design of the Tetris Machine

The design is entered using a drawing program `xdp/wireC` schematic drawing tool and is translated to the Xilinx XNF format by `xnfwirec` [2]. These tools combine the benefits of a drawing tool and procedural language, and are well integrated with the synthesis tools that we described in earlier chapters. The reader is invited to read [2] for details. `Xdp/wirec` is really not much different from commercial tools, except that it has more capabilities and is run under the X-window environment.

Figure 9.20 shows the top-level partitioning of the Tetris machine. The machine is divided into two main blocks: the controller `cont` and the data path `data`. Figure 9.20 shows all the I/O pads (`P`), buffers (`IB`,`OB`) of the two

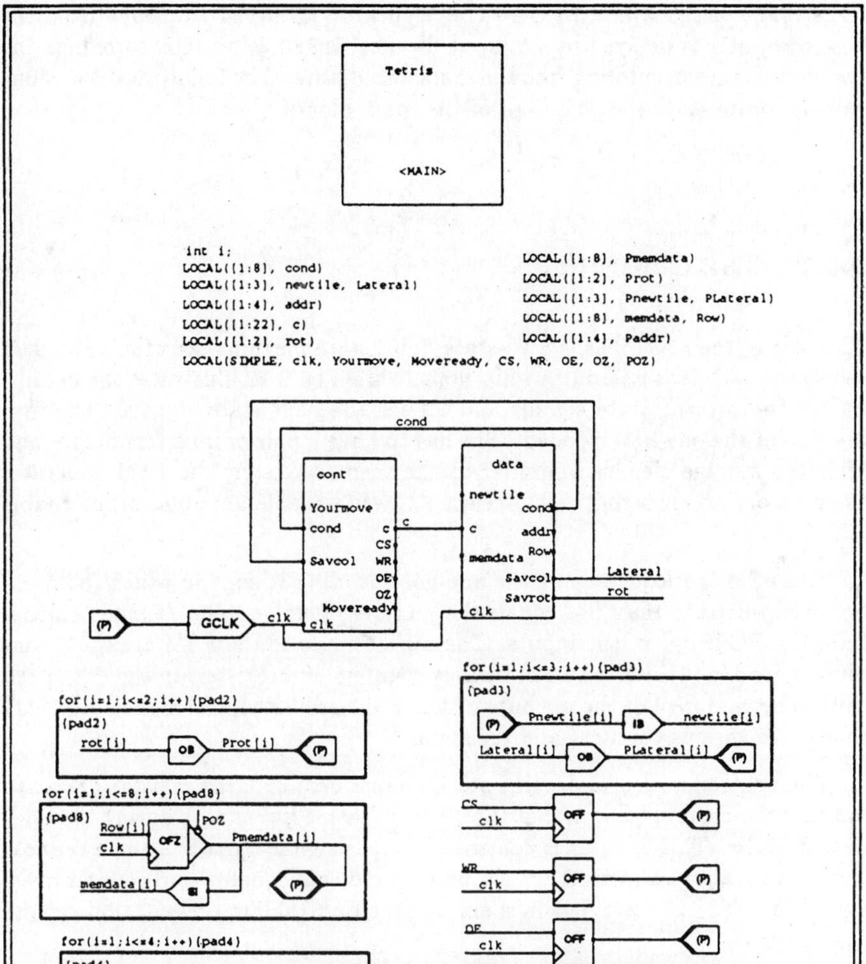

Figure 9.20: xnfwirec Tetris Design, Top-Level Diagram.

main blocks, and the interface signals with the host computer. Notice that the memory signals CS, WR, and OE are "deglitched" by the flip-flops at the IOBs (OFF) of the Xilinx FPGA. The asynchronous input Yourmove from the host computer is delayed by an input flip-flop INFFQ at an IOB to reduce the system's chance of entering into the metastable state. The following discussions will elaborate on the logic design of the main blocks.

9.5.1 The Controller

The core of the controller is a 22-state finite state machine contfsm. Fig. 9.21 shows the I/O pads and output flip-flops, while Fig. 9.22 illustrates the details. PS are the present state signals and NS are the next state signals. The first flip-flop of the one-hot-encoded FSM has to have a pair of inverters before and after the storage element. Similar to the memory tester, the FSM generates the memory access signals: chip-select CS, write-enable WR, and output-enable OE.

The FSM is encoded using the one-hot encoding scheme in which the number of flip-flops in the FSM equals the number of states. The Yourmove signal from the PC is one of the inputs. The other inputs to the FSM are the 8 conditions cond[1:8] derived from various counters and registers in the data path chip. The majority of the outputs of the FSM are 22 control signals c[1:22] that drive various counters and registers.

The controller also maintains a 3-bit *increment* or *clear* register ctcol to indicate the *current* column that the Tetris machine is scanning during the evaluation phase. This register is controlled by the signals incctol and clrctol. Two signals eqcol and fullcol are used to indicate equality and counter overflow. The FSM is described in a state transition table representation on the next page.

Referring to Fig. 9.23, the initialization sequence starts with states S0 to S2, which clears the first 16 bytes of the SRAM to reflect the status of an empty Tetris bucket of 16 rows by 8 columns. States S3 and S4 are used to poll low assertion of the signal Yourmove by the host PC. When this happens, the machine starts to evaluate the best placement of the new tile. The machine then produces the necessary row masks in states S6 to S8 and updates the memory in states S11 to S13. It goes back to state S6 from S12 because the preceding updating step has to be repeated 16 times representing the number of rows that the machine is keeping track of. States S14 to S15 are there to update the counters Lrow (last row) and Crow (current row); the functions of these counters will be explained in the next subsection. You probably have the sense that the machine is really spending most of its effort in maintaining the status of the board.

```
.i 11
.o 29
.s 22
.ilb eqcol fullcol Yourmove cond[1] cond[2] cond[3] cond[4] cond[5]
     cond[6] cond[7] cond[8] PS[1] PS[2] PS[3] PS[4] PS[5] PS[6]
     PS[7] PS[8] PS[9] PS[10] PS[11] PS[12] PS[13] PS[14] PS[15]
     PS[16] PS[17] PS[18] PS[19] PS[20] PS[21] PS[22]
.ob  NS[1] NS[2] NS[3] NS[4] NS[5] NS[6] NS[7] NS[8] NS[9] NS[10]
     NS[11] NS[12] NS[13] NS[14] NS[15] NS[16] NS[17] NS[18] NS[19]
     NS[20] NS[21] NS[22]
     CS WR OE OZ c[1] c[2] c[3] c[4] c[5] c[6] c[7] c[8] c[9] c[10]
     c[11] c[12] c[13] c[14] c[15] c[16] c[17] c[18] c[19] c[20]
     c[21] c[22] incctcol clrctcol Moveready
#inputs: eqcol fullcol Yourmove cond1 cond2 cond3 cond4 cond5 cond6
#        cond7 cond8
#outputs: CS WR OE OZ c1 thru 22 incctcol clrctcol Moveready
#                                    1111111111222
#fY12345678                 12345678901234567890l2icM
----------- S0  S1  0110--0000-00----0-0-----1--1
----------- S1  S2  0010--0000-00----0-0-----1--1
------1---- S2  S3  1110--1000-00----0-0-1-101--1
--1-------- S3  S3  1111--00---00----0-00--001--0
--0-------- S3  S4  11110101---00----0-00--11---1
--0-------- S4  S4  11110000---00----0-00--0----1
--1-------- S4  S5  11110000--100010111100110---1
----0------ S5  S5  11110000--00000100-01--0----1
----11----- S5  S6  11110101--00101010-01--0--011
0-------10- S6  S7  11110000--00010000-001-10-101
0--------1- S6  S7  11110000--00010000-001-10-101
1-------10- S6  S7  11110000--00010100-001-10-101
1--------1- S6  S7  11110000--00010100-001-10-101
--------00- S6  S8  11110000--00000000-001-101001
00--0-----0 S7  S7  11110000--00000000-000010-101
10--------0 S7  S7  11110000--00000100-000010-101
-0--------1 S7  S7  11110000--00000100-000010-101
-0--1------ S7  S7  11110000--00000100-000010-101
01--0-----0 S7  S8  11110000--00000000-0000101101
11--------0 S7  S8  11110000--00000100-0000101101
-1--------1 S7  S8  11110000--00000100-0000101101
-1--1------ S7  S8  11110000--00000100-0000101101
0---0-----0 S8  S9  01010000--00000000-0000101001
1---------0 S8  S9  01010000--00000100-0000101001
----------1 S8  S9  01010000--00000100-0000101001
----1------ S8  S9  01010000--00000100-0000101001
---1--0---- S10 S6  1111001000000000O-00--0-0001
---1--1---- S10 S13 1111001001000----0-00----0--1
---0------- S10 S11 01100000000000000-00--0-0001
------0---- S12 S6  11111010--00000000-00--0--001
------10--- S12 S13 1111101001000----0-00-------1
```

```
------11--- S12 S3  1111101001000----0-001-10---1
----------- S13 S14 0110000000-00----0-00-------1
----------- S14 S15 0010000000-00----0-00----0--1
-------1--- S15 S3  11101000---00----0-001-100--1
-------0--- S15 S13 1110100000-00----0-00----0--1
----------- S16 S17 11110000000000000-00--0-1001
----------- S17 S10 11110000000000000-00--0-0001
----------- S11 S18 00100000000000000-00--0-0001
----------- S18 S12 11100000001000000-00--0-0001
------0---- S2  S19 1110--1000-00----0-0-----1--1
----------- S19 S1  0110--0000-00----0-0-1-101--1
----------- S9  S20 01010000000000010-000010100 1
----------- S20 S16 01010000100000010-00--0-1001
----10----- S5  S21 11110000--0001010100101 10---1
----------- S21 S5  11110000--0000000000--0----1
```

Exercise 4: In the Tetris FSM, the Boolean expressions for the $Moveready$ signal as generated by **mustang** and **misII** is given in (9.3). Suggest a simple way to further simplify the signal $Moveready$.

$$
\begin{aligned}
Moveready &= [66] + [63] + [62] + [149] \\
[149] &= cond4\ cond5\ PS14 + cond5\ PS16 + cond4\ PS3 \\
[62] &= PS21 + PS10 + PS9 \\
[63] &= [68] + PS15 + PS13 + PS2 \\
[68] &= NS13 + NS2 + PS12 \\
[66] &= [148] + NS20 + NS12 + PS19 \\
[148] &= NS22 + NS9 + NS8 + NS7 \\
&\quad + NS6 + NS5 + PS18 + PS17
\end{aligned} \tag{9.3}
$$

9.5.2 The Data Path

The **Data** block is the *evaluation function* and the bookkeeper of the Tetris machine. The evaluation function scans all eight columns of the Tetris bucket and tries all four rotations of the tile at each column to locate the best position to land the tile. The image of the Tetris bucket in the RAM is then updated. The evaluation function and updating logic are the most involving and interesting parts of the Tetris machine.

Figure 9.24 shows all the I/O pads of the data path chip, and Fig. 9.25 shows all the registers and counters. The submodules in this data path chip are primarily for maintaining and updating internally the status of the Tetris bucket, so that the Tetris machine can correctly judge the best location (column and row) and rotation to place the new tile. The submodules are not for

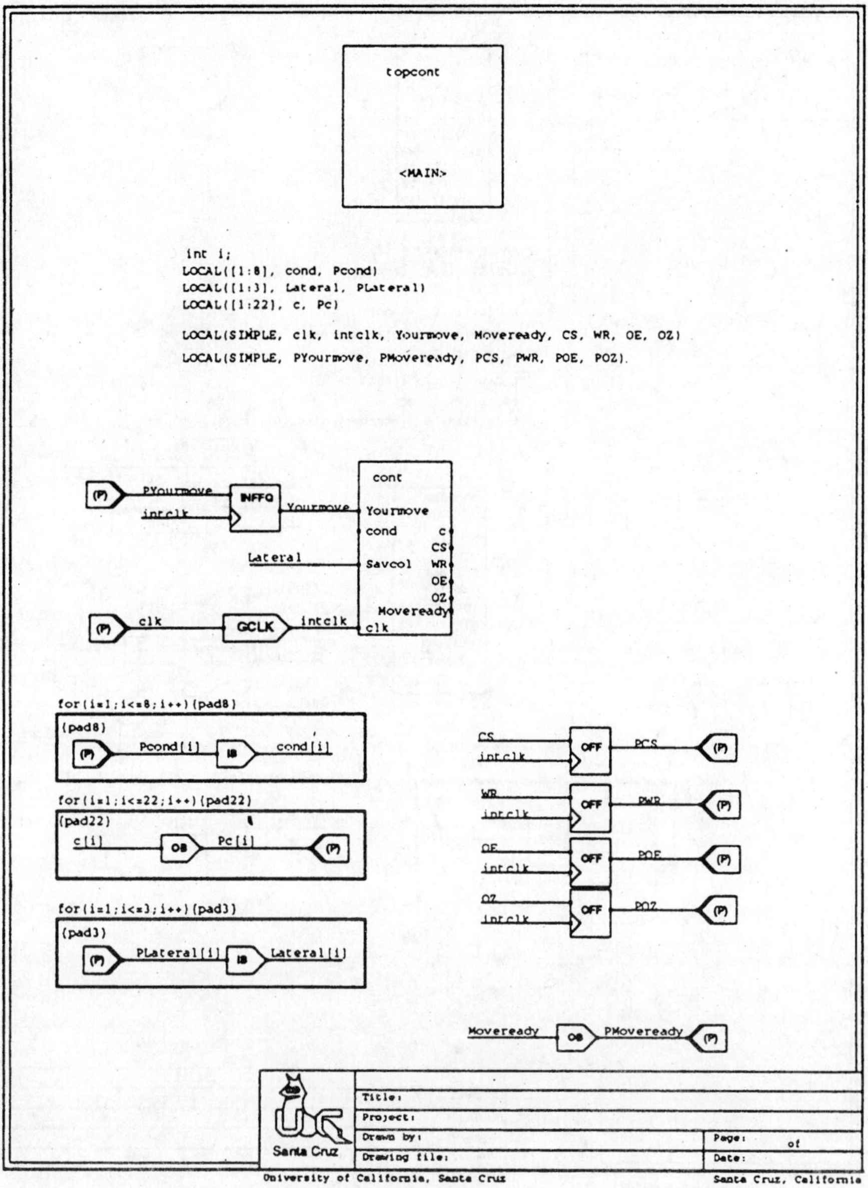

Figure 9.21: The Controller of the Tetris Machine, Showing the I/O Pads and the FSM.

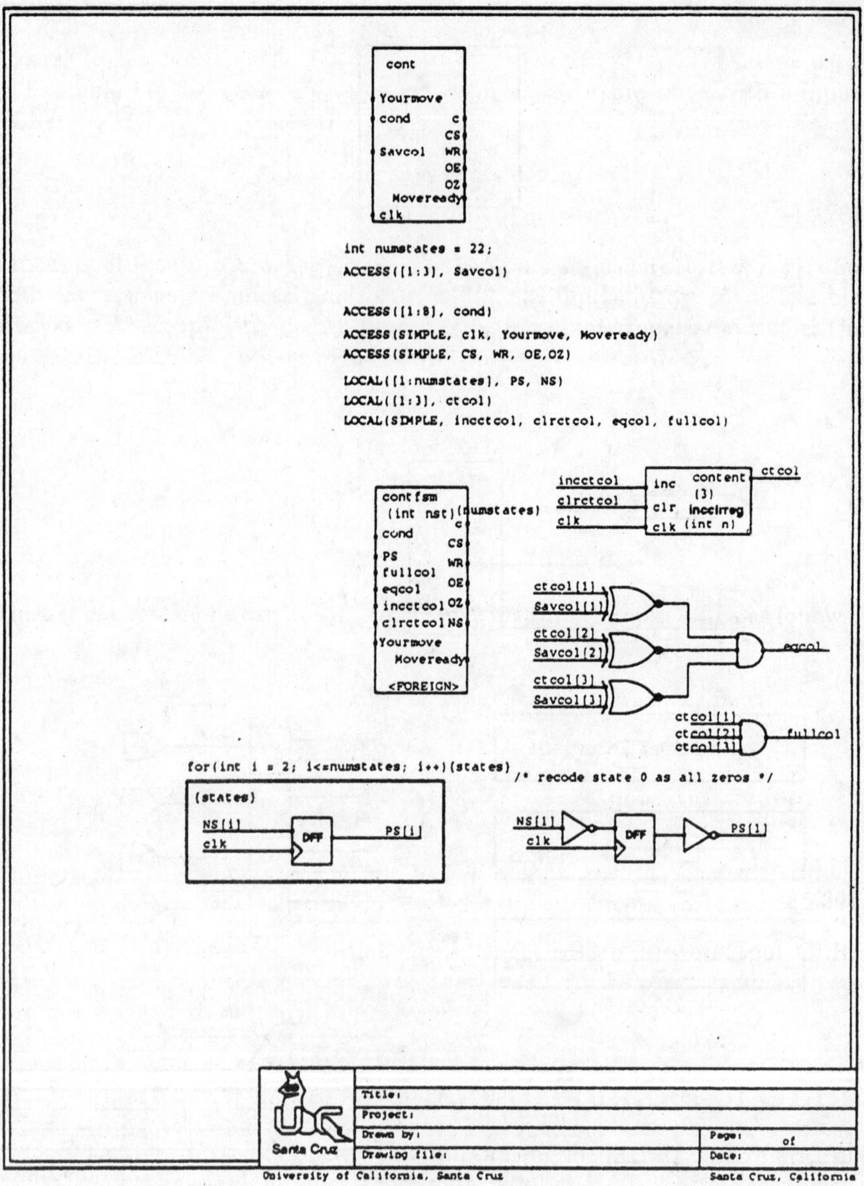

Figure 9.22: The FSM of the Controller.

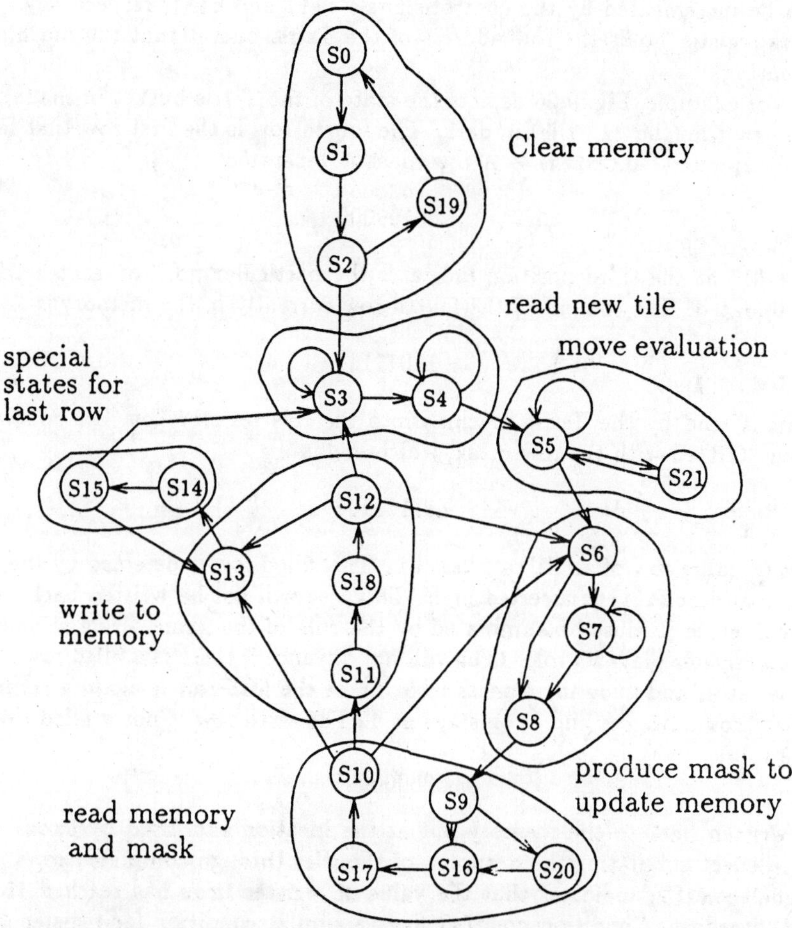

Figure 9.23: State Diagram of the FSM of the Tetris Machine.

displaying the status of the Tetris bucket. The functions of the registers and counters deserve some detail explanation.

The updating of the memory is primarily performed by two row registers and a row mask. At initialization, the `Lrow` (last row) register and `Crow` (current row) registers are cleared. These two registers are 4 bits each, corresponding to the 16 (out of 20) rows that the machine is keeping track of. They can be incremented by the control signals `c[1]` and `c[3]`, respectively. The `Crow` register holds the row address of the Tetris bucket that the machine is updating.

For example, Fig. 9.26 depicts the state of the Tetris bucket in shade, and the new tile (the "T" tile) in dark. The fourth row is the first row that needs to be updated, so `Crow` is 4. A row mask is generated

00100000

The "1" at the third position indicates the particular position that needs to be updated. The content of the fourth row currently in the memory is

11011111

This is read by the Tetris machine into the row register `Row`. Its content is then "OR"ed with the row mask, which yields

11111111

Let us call a row of all "1" a *filled row*. This filled row is detected by the AND gates and `cond[1]` is asserted high. This row will not be written back to the RAM, since it must be eliminated by the rule of the game. Here is how the `Lrow` register plays its role. `Lrow` will not advance if there is a filled row. `Lrow` is set at 4, and `Crow` increments to 5. Since the fifth row is again a complete row, `Crow` advances and `Lrow` stays at 4. The sixth row is not a filled row, so the value

00000001

is written back to the memory but at the location addressed by `Lrow`. This is in effect simulating the dropping of the tiles through completed rows. The signal `cond[5]` indicates that the value of register `Lrow` has reached 15 and will overflow. Condition `cond[4]` flags a similar condition for register `Crow`. These conditions indicate the completion of the memory update cycle.

The rest of the submodules in Fig. 9.25 perform the following functions.

1. `H` registers: there are eight 4-bit parallel-load height registers that hold the maximum height (skyline) of each column in the Tetris bucket. As we shall see later, only the skyline of the Tetris bucket is needed during move evaluation. The outputs of the registers drive a shared bus. These height registers are loaded from an 8-bit register `Row`, which buffer the data fetched from the memory. Since the image of the Tetris bucket is

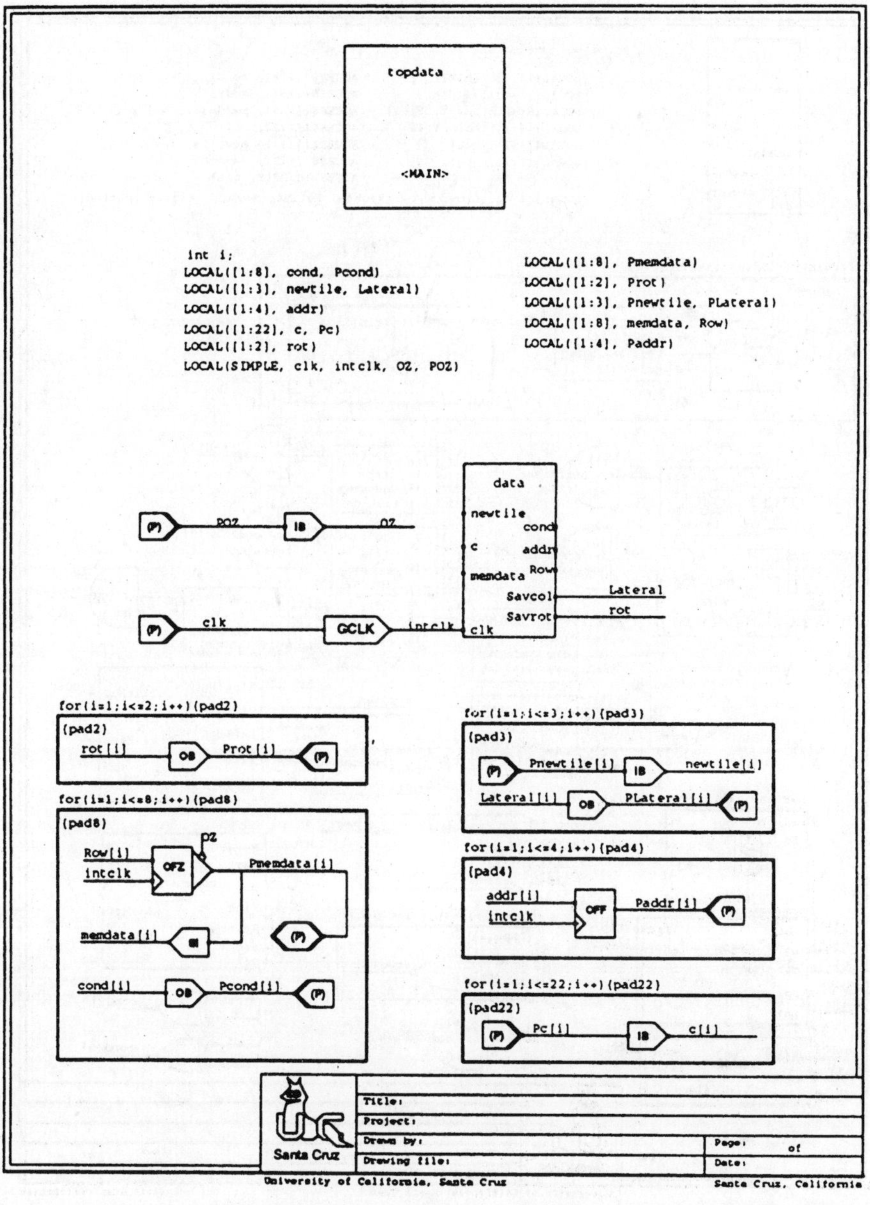

Figure 9.24: Data Path Chip, Top-Level Diagram.

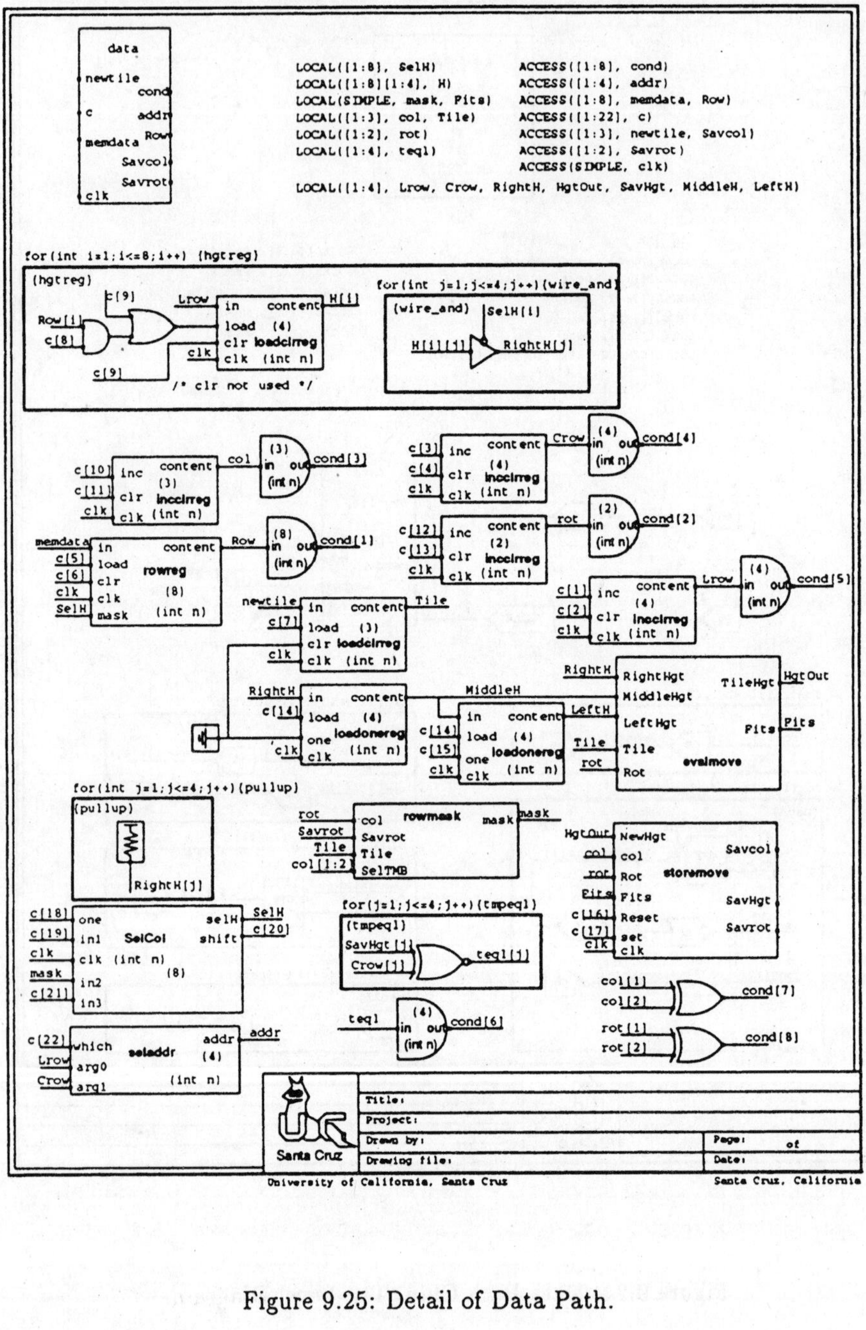

Figure 9.25: Detail of Data Path.

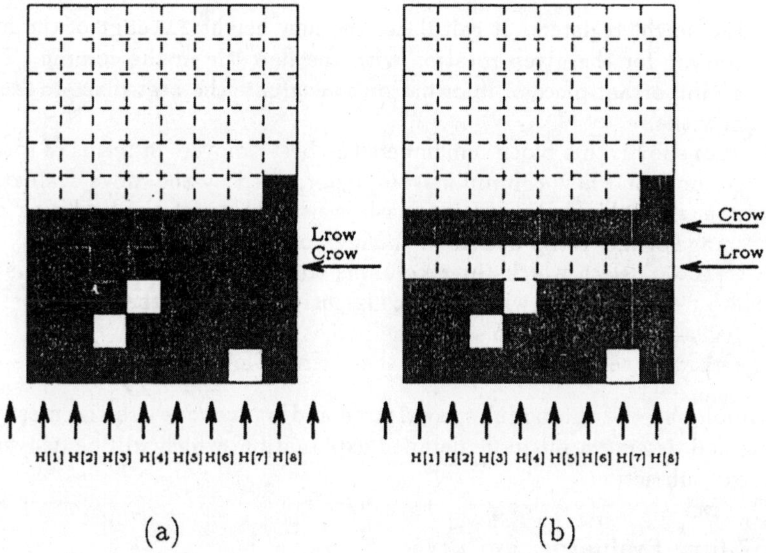

Figure 9.26: Tetris Bucket Illustrating the `Crow`, `Lrow` and the Height Registers.

stored rowwise in the memory, and the `H` registers hold the maximum height of each column in the Tetris bucket, we need to convert rowwise information to column heights. If `Row[i]` is "1," then the content of the `Lrow` register is loaded into the height register `H[i]`. All 16 rows are scanned by incrementing the `Lrow` register from 0 to 15, during which the memory is also updated.

2. `Row` register: this 8-bit parallel-load register holds the current value fetched from the memory. The values are fetched one row (one byte) at a time. When all 8 bits in the row are ones, `cond[1]` will be asserted.
3. `Tile`: this parallel-load 3-bit counter keeps the encoding of the new tile from the host computer.
4. `col`: this increment or clear 3-bit counter keeps the address of the best column that the new tile will land.
5. `rot`: this increment or clear 2-bit counter is used to step through all four possible rotations of the tiles.
6. `RightH`: this is a 4-bit-wide shared bus; the contents are coming from the height registers. The height registers are selected by an 8-bit shift register `SelH` having three consecutive "1"s. `LeftH`, `MiddleH`, and `RightH` form a set of registers that holds the heights of the three adjacent columns during the move evaluation. Moves are evaluated based primarily on the values of these registers, the rotation, and the new tile.
7. `evalmove`: this is the *move evaluator* of the Tetris machine. It takes the tile type, the degree of rotation, and three adjacent column heights to determine whether there is a perfect fit between the new tile and

the height registers. It calculates the new height `TileHgt` of the middle column for the given rotation with the new tile in the column. This is an important piece of information to evaluate the best place to land the new tile.

8. `storemove`: this block remembers the best "move," in terms of rotation, column, and new column heights, determined by the move evaluator.
9. `rowmask`: this 8-bit register mask is used to assist the updating of the image of the Tetris bucket in the memory.
10. `SelCol`: select which three columns to evaluate. This is essentially an 8-bit shift register with three consecutive ones and the rest zeros. This register is also used to select the height registers.
11. `seladdr`: select the address of the memory locations.

Among all the submodules, `evalmove` and `storemove` are the most interesting and deserve a bit more detailed explanation, which will be delivered in the next subsection.

The Move Evaluator: `evalmove`

This "move" evaluator takes the tile type, the degree of rotation, three adjacent column heights, and determines whether there is a fit between the new tile and the terrain of the bucket in these columns. This combinational block produces the new height of the middle column at which the tile will land on, with or without a perfect fit. The score of a move depends on the new height and whether the tile fits the terrain for the given rotation. The best score so far is kept in the `storemove` block of the Tetris machine.

The `evalmove` is decomposed into seven submodules so as to exploit the five-input look-up table architecture of a Xilinx XC3000 FPGA. We did try to implement this function without manual decomposition, but the synthesis tool was not able to exploit the structure of the function and yielded prohibitive results. Referring to Fig. 9.27, the subblock `eval` is a seven-output combinational function. The inputs to this function `eval` is the tile type in 3 bits, and a given rotation in 2 bits, all together 5 bits. The outputs are `incL`, `incR`, `incM`, `inc2M`, `sub1`, `useR`, and `useL`, which are used to control the updating of the new height. Each output uses one five-input look-up table. As shown in Fig. 9.27, the Boolean function `eval` is a decoder that controls the rest of the modules in `evalmove`: the maximum function `max`, incrementer `inczero`, and the subtracter `minus1`. The function of `eval` is specified in the `bdsyn` high-level language, which was discussed earlier in Chapter 3. The specification of the function is listed in Figs. 9.28 and 9.29, which essentially describes a decoder.

Given a rotation, the `evalmove` function classifies the new tile into 15 groups based on the outcome of the new column height `TileHgt`. This information will be used to determine the overall best column to place the new tile. The new height is measured with respect to the center of the tile (see Fig. 9.17). For example, there are only four possible cases for any tile to span three consecutive columns with the center of the tile in the middle; they are

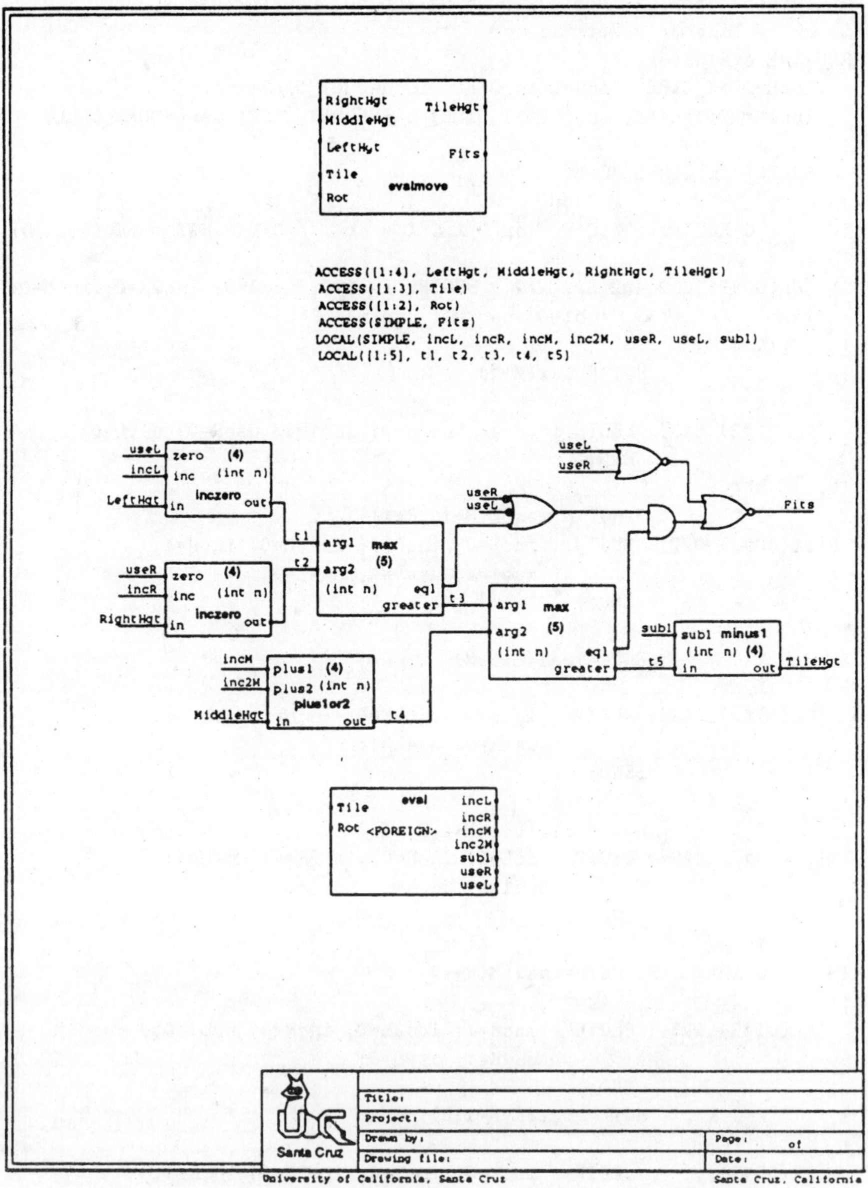

Figure 9.27: Move Evaluator of the Tetris Machine.

```
MODEL evalmove
   incL<0>,incR<0>,incM<0>,inc2M<0>,sub1<0>,useR<0>,useL<0>
      = Tile<3:1>,Rot<2:1>;
ROUTINE evaluator;
   incL=DONT_CARE; incR=DONT_CARE; incM=DONT_CARE;
   inc2M=DONT_CARE; sub1=DONT_CARE; useR=DONT_CARE; useL=DONT_CARE;

   SELECT Tile&Rot FROM
!                   X                X           X
!     X C X    or   X C X   or   X C X   or   X C X   HGT = max(L,M,R)
!
   [6&(00#2),6&2,4&2,5&2,2&2]: BEGIN  incL=0; incR=0; inc2M=0; incM=0;
                               sub1=0; useR=1; useL=1; END;
!      X
!      C          HGT = max(M+1)
!      X
  [6&(01#2),6&3]: BEGIN incM=1; inc2M=0; sub1=0; useR=0; useL=0;
                  END;
!      X X
!      C X          HGT = max(L,M+1, R+1)-1
  [0&(00#2),0&2]: BEGIN    incL=0; incR=1; inc2M=0; incM=1;
                           sub1=1; useR=1; useL=1;
                  END;
!        X
!      X C        HGT = max(L+1,M)
!      X
  [0&(01#2),0&3]: BEGIN    incL=1; inc2M=0; incM=0; sub1=0;
                           useR=0; useL=1;
                  END;
!        X X
!      X C        HGT = max(L+1,M+1,R)-1
  [1&(00#2),1&2]: BEGIN  incL=1; incR=0; inc2M=0; incM=1;
                         sub1=1; useR=1; useL=1;
                  END;
!      X
!      C X         HGT = max(M,R+1)
!        X
  [1&(01#2),1&3]: BEGIN    incR=1; inc2M=0; incM=0; sub1=0;
                           useR=1; useL=0;
                  END;
!      X C X     HGT = max(L,M+1,R)
!        X
  [2&(00#2)]:     BEGIN    incL=0; incR=0; inc2M=0; incM=1;
                           sub1=0; useR=1; useL=1;
                  END;
!      X
!      C X         HGT = max(M+1,R)
!      X
  [2&(01#2)]:     BEGIN    incR=0; inc2M=0; incM=1; sub1=0;
                           useR=1; useL=0; END;
```

Figure 9.28: Bdsyn Description of Eval (Part I).

```
!       X
!     X C          HGT = max(L,M+1)
!       X
  [2&3]:        BEGIN  incL=0; inc2M=0; incM=1; sub1=0;
                       useR=0; useL=1;
                END;
!     X X
!     X C          HGT = max(L,M)
  [3&(00#2),3&(01#2),3&2,3&3]:
                BEGIN  incL=0; inc2M=0; incM=0; sub1=0;
                       useR=0; useL=1;
                END;
!     X C X        HGT = max(L+1,M,R)
!     X
  [4&(00#2)]:   BEGIN   incL=1; incR=0; inc2M=0; incM=0;
                        sub1=0; useR=1; useL=1;
                END;
!     X
!     C            HGT = max(M+1,R+1)
!     X X
  [4&(01#2)]:   BEGIN   incR=1; inc2M=0; incM=1; sub1=0; useR=1;
                        useL=0;
                END;
!     X X
!       C          HGT = max(L,M+2)-1
!       X
  [4&3]:        BEGIN  incL=0; inc2M=1; incM=0; sub1=1; useR=0;
                       useL=1;
                END;
!     X C X        HGT = max(L,M,R+1)
!         X
  [5&(00#2)]:   BEGIN  incL=0; incR=1; inc2M=0; incM=0; sub1=0;
                       useR=1; useL=1;
                END;
!     X X
!     C            HGT = max(M+2,R)-1
!     X
  [5&(01#2)]:   BEGIN  incR=0; inc2M=1; incM=0; sub1=1; useR=1;
                       useL=0;
                END;
!       X
!       C          HGT = max(L+1,M+1)
!     X X
  [5&3]:        BEGIN   incL=1; inc2M=0; incM=1; sub1=0; useL=1;
                END;
  ENDSELECT;
ENDROUTINE evaluator;
ENDMODEL;
```

Figure 9.29: Bdsyn Description of Eval (Part II).

```
              X            X           X
 X C X      X C X        X C X       X C X
  (1)        (2)          (3)         (4)
```

where the character `C` denotes the center of the tile. The first case happens with the new tile code "6" and the rotation code is "2." See Fig. 9.17 for the encoding of the tiles and rotations.

In any of these cases, the new height less one must be

$$\max(left\_height, middle\_height, right\_height)$$

since the tile has to span three columns and there is a perfect fit if the heights of the three adjacent columns are all equal. We need new height less one instead of the actual new height since this is the row at which we should start to update the memory. The seven output control signals of the `eval` function select the proper way to use the two comparators (`max`) to compare the heights.

As another example, the "bar" tile number 6 if it stands upright, the new height must be

$$middle\_height + 1$$

So the output `incM` of this function is activated to increment the value of the middle-column height registers with the "plus1or2" function, while the left and right values are ignored.

The `eval` specification is translated into Boolean equations by the `bdsyn` translator. It is further minimized by the multilevel logic minimizer `misII`. The minimized equations, which can be implemented in seven CLBs is given as follows.

$$
\begin{aligned}
INORDER &= Tile3\ Tile2\ Tile1\ Rot2\ Rot1; \\
OUTORDER &= incL\ incR\ incM\ inc2M\ sub1\ useR\ useL; \\
incL &= Tile3\ Tile1\ Rot2\ Rot1 + Tile3\ Tile2'\ Tile1'\ Rot2' \\
&\quad + Tile3'\ Tile2'\ Rot1 + Tile3'Tile2'\ Tile1 \\
incR &= Tile3\ Tile1\ Rot2'\ Rot1' + Tile2'\ Tile1'Rot2'\ Rot1 \\
&\quad + Tile3'\ Tile2'\ Rot1 + Tile3'\ Tile2'\ Tile1' \\
incM &= Tile3'\ Tile1'\ Rot2'\ Rot1' + Tile3\ Tile1'\ Rot2'\ Rot1 \\
&\quad + Tile3\ Tile1\ Rot2\ Rot1 + Tile3'\ Tile2'\ Rot1' \\
&\quad + Tile2\ Tile1'\ Rot1 \\
inc2M &= Tile3\ Tile2'\ Tile1'Rot2\ Rot1 + Tile3\ Tile1\ Rot2'\ Rot1 \\
sub1 &= Tile3\ Tile2'\ Tile1'\ Rot2\ Rot1 + Tile3\ Tile1\ Rot2'\ Rot1 \\
&\quad + Tile3'\ Tile2'\ Rot1' \\
useR &= Tile3\ Tile1\ Rot2'\ Rot1 + Tile3\ Tile2'\ Tile1'\ Rot2' \\
&\quad + Tile3'\ Tile2\ Tile1'\ Rot2' + Tile3'\ Tile2'\ Tile1 \\
&\quad + Tile1'\ Rot1' + Tile2'\ Rot1' \\
useL &= Tile3\ Tile2'\ Rot2 + Tile3'\ Tile2\ Rot2 + Tile3'\ Tile2'\ Tile1' \\
&\quad + Tile2\ Tile1 + Rot1'
\end{aligned}
$$

Keeping the Best Move: The storemove Module

So the evaluation function scans all eight columns of the Tetris bucket and tries all four rotations of the tile at each column. The best move is kept by the **storemove** submodule, which keeps track of the rotation, and column, which provides the best move. The moves are scored based on the new height **NewHgt** (generated by the **evalmove** submodule) and whether the tile fits. The new height and the "fit" bit are bundled and are fed to a comparator in **storemove**, as depicted in Fig. 9.30. The score of a move is compared with the best score recorded so far by the comparator, and the move of the winner will be saved into the registers. Since scanning always starts from the left (column 1), this raises the issue of balance in the event that there is more than one best move. Otherwise, tiles will start to pile up on one side. The toggle flip-flop on the right provides the mechanism for alternating the tie-breaking rule after each scan.

Updating the Tetris Bucket Status: rowmask

The image of the Tetris bucket in the memory is updated by reading one row (8 bits) of the bucket at a time and writing the 8 bits back to the memory through a row mask **SelH** in the row register **rowreg**, as illustrated in Fig. 9.26. A 4-bit row counter **Crow** starts from zeros, and keeps incrementing until it reaches the same value as **SavHgt** kept in **storemove**. Then a 1-bit mask is created by the **rowmask** submodule. This mask is shifted out 1 bit at a time into the **SelCol** submodule, which generates the complete mask **SelH** after eight cycles. Since a tile is at most 3×3 in size, the 2-bit **SelTMB** input of this submodule determines which "row" (top, middle or bottom) of the tile the 1-bit mask is being applied to, as depicted in Fig. 9.32. Similarly, a 2-bit input **col** of this submodule determines the column position of the 1-bit mask. The encoding of a tile is **mask(row,col)** as given:

```
              7 8 9
      row     4 5 6
              1 2 3

               col
```

The **decodetile** Boolean function inside the **rowmask** submodule serves the function of decoding and generating the bit map of a tile. This Boolean function is specified in the **bdsyn** language as given in Fig. 9.31. The variable **dectile** is the bit map of a tile. A 1-bit mask **mask** is generated according to the row **SelTMB** and column **col** selected.

This Tetris machine used 116 CLBs (out of 144 available) in the data path chip and 58 CLBs (out of 64 available) in the controller chip. The machine runs at about 5 MHz using the XC3000-50 speed grade parts.

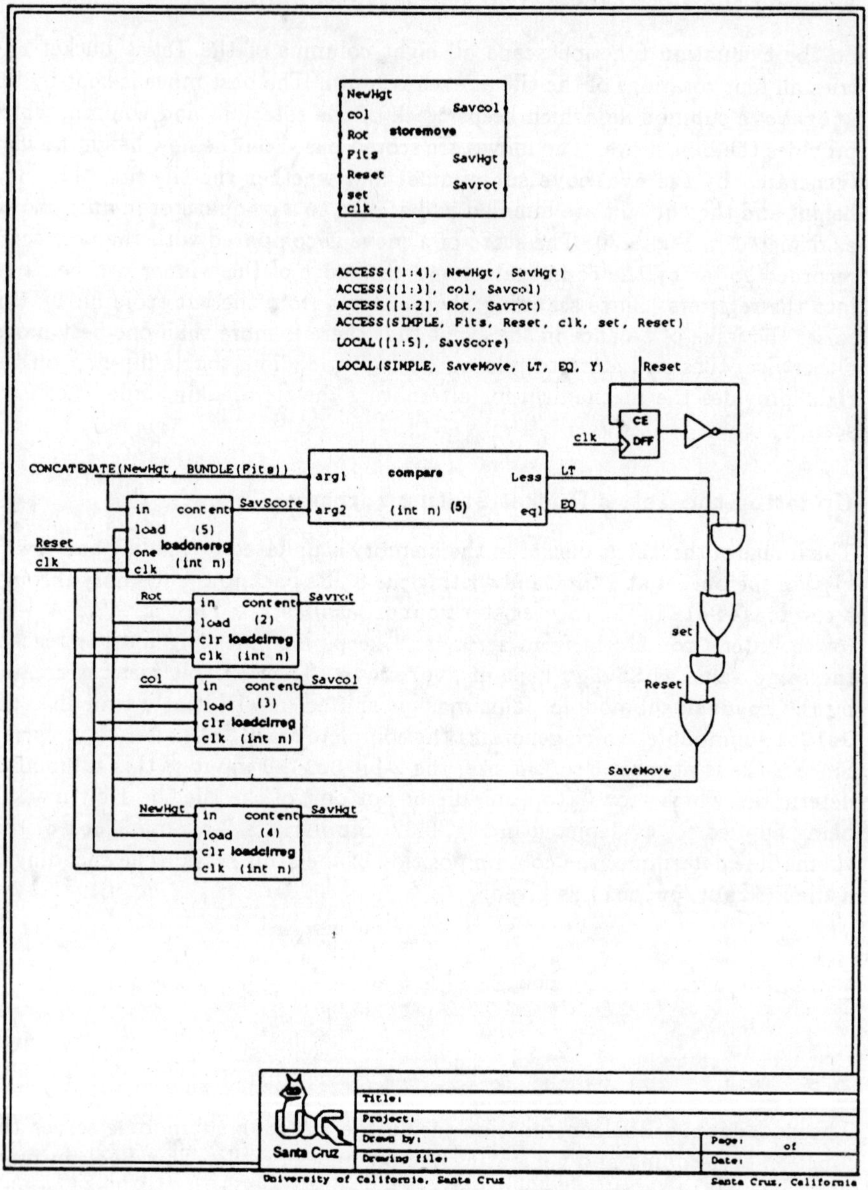

Figure 9.30: Tetris Score and Move Keeper.

```
MODEL decodetile
      dectile<9:1> = Savrot<2:1>, Tile<3:1>;

ROUTINE decode;
! position 5 is always on
   dectile = 16;                                      !   7 8 9
                                                      !   4 5 6
 SELECTALL Tile&Savrot FROM                           !   1 2 3
 [0&(01#2),0&3,5&3,4&(00#2)]:BEGIN
                dectile<1> = 1;
           END;
 [6&(01#2),6&3,4&3,5&(01#2),2&(00#2),5&3,4&(01#2),2&(01#2),2&3]:BEGIN
                dectile<2> = 1;
           END;
 [1&(01#2),1&3,5&(00#2),4&(01#2)]:BEGIN
                dectile<3> = 1;
           END;
 [6&(00#2),6&2,0&(01#2),0&3,1&(00#2),1&2,3&(00#2),3&(01#2),3&2]:BEGIN
                dectile<4> = 1;
           END;
 [3&3,4&2,5&2,2&2,5&(00#2),4&(00#2),2&(00#2),2&3]:BEGIN
                dectile<4> = 1;
           END;
 [6&(00#2),6&2,0&(00#2),0&2,1&(01#2),1&3,4&2]:BEGIN
                dectile<6> = 1;
           END;
 [5&2,2&2,5&(00#2),4&(00#2),2&(00#2),2&(01#2)]:BEGIN
                dectile<6> = 1;
           END;
 [0&(00#2),0&2,3&(00#2),3&(01#2),3&2,3&3,5&2,4&3]:BEGIN
                dectile<7> = 1;
           END;
 [6&(01#2),6&3,0&(00#2),0&2,0&(01#2),0&3,1&(00#2),1&2,1&(01#2),1&3]:
           BEGIN
                dectile<8> = 1;
           END;
 [3&(00#2),3&(01#2),3&2,3&3,4&3,5&(01#2),2&2,5&3,4&(01#2),2&(01#2),2&3]:
           BEGIN
                dectile<8> = 1;
           END;
 [1&(00#2),1&2,5&(01#2),4&2]:BEGIN
                dectile<9> = 1;
           END;
 ENDSELECTALL;
ENDROUTINE;
 ENDMODEL;
```

Figure 9.31: Tile Decoder Specified in the **Bdsyn** Language.

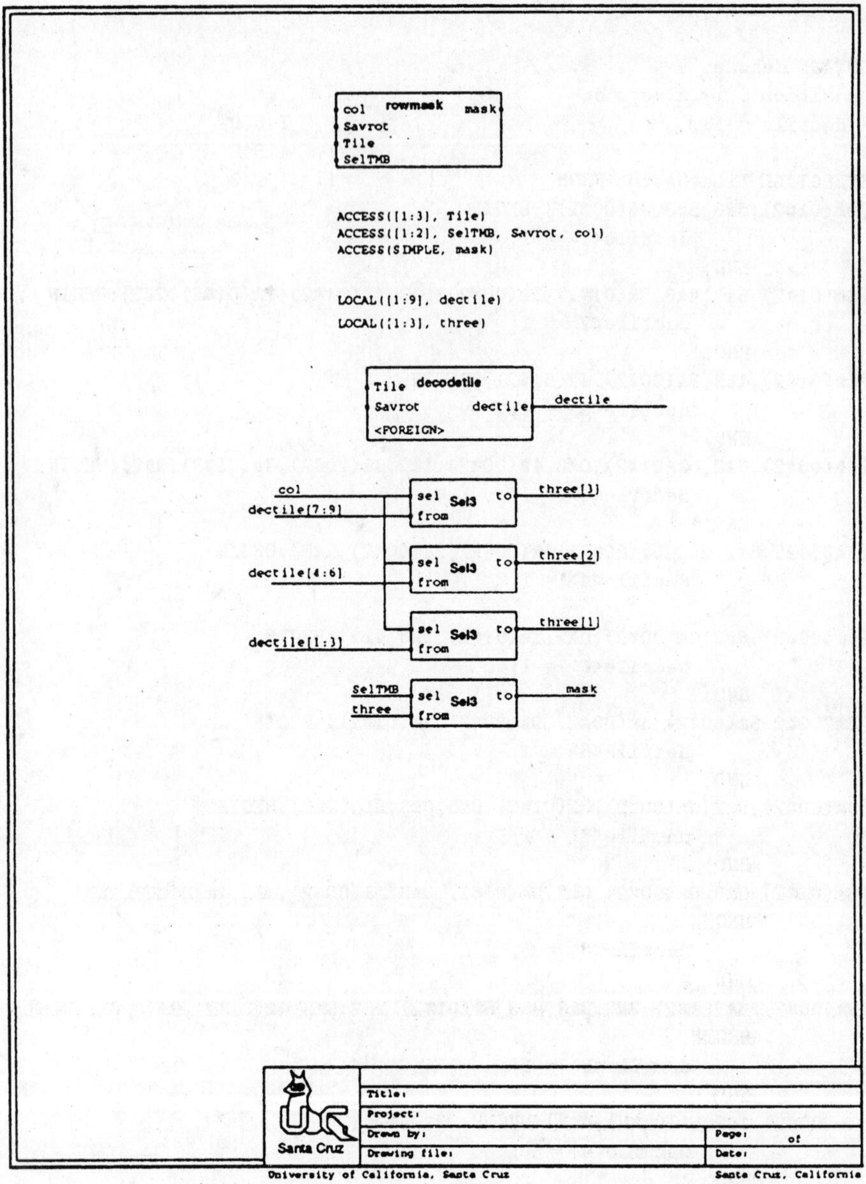

Figure 9.32: Row Mask for Updating Bucket Status in Memory.

Bibliography

[1] XILINX: *The Programmable Gate Array Data Book.* 2100 Logic Drive, San Jose, CA 95124, 1993.

[2] Kong, J., M. Schlag, and P. K. Chan. XNFWIREC tutorial: the Wirec manual derivative. Technical report, University of California, Santa Cruz, June 1991.

[3] Schlag, M., J. Kong, and P. K. Chan. Routability-driven technology mapping for lookup table-based FPGAs. In *1992 IEEE International Conference on Computer Design: VLSI in Computers and Processors*, pages 86–90, Cambridge, Massachusetts, Oct. 1992.

Appendix A

The Tetris PC Host Program

A.1 Appendix: Tetris Driver

A listing of the PC tetris host is given here. This program must be compiled with `IBMPC` defined, for example,

```
tcc -DIBMPC host.c
```

using Borland Turbo C. This program assumes that your computer has the `ansi.sys` driver installed.

```
/*********************************************************
  Name: host.c
  Tetris PC host

  Modify from a Tetris code in the public domain
  Credit to the unknown original author

  Spring 1991,     University of California, Santa Cruz
                   Pak K. Chan 5/17/91
 *********************************************************/
#include<stdio.h>
#include<time.h>
#include<stdlib.h>

#ifdef IBMPC
  #include<dos.h>
  #include<conio.h>
  #include<bios.h>
  unsigned int PortD, PortJ, PortK;
  void YourMove(), set_up(), get_move(), outportbbb();
```

```
  int  ReadMove();
  unsigned int inportbbb();
  char * port_name();
  char M[99];     /* character arrays for player's moves */
#endif

#define GAMES 999
int Scores[GAMES], ngames;
float total=0;
int   even=0;

#define THINKTIME  1000
#define SLEEPTIME  5
#define SHAPES   4
#define TYPES    7  /* the number of different tiles 7 max*/
#define X        10
#define XM1      (X/2)
#define XSTART   X+(X-2)/2+1
#define X2       2*X+1
#define Y        20
#define XY       X*(Y+1)
#define FRAME    X*(Y+3)
#define BUCKET   X*(Y+2)
#define is        ==
#define isnt      !=

/*
  The tetris bucket is of size 10 by 23  = 230

  f[] is the  main data structure of the tiles
  there are 7 different tiles: the first 7 entry in f[].
  Each entry is of 4 integers.
  The first integer points to the rotated tile, the
  next 3 integers are the coordinates of the rest of the shape.
  Example:
     7, -13,12,1 is the "z" tile

        -13 -12
            +00 +01

  when it rotates it becomes

            -12
        -01 +00
        +11

  The function u() updates the screen.
  The for loop in the function main picks a tile at random,
  and determines whether the player has stuck a key (c<0);
```

```
    if not, it produces the effect that the tile is dropping down.
    check whether there are any complete rows,
    and update the screen.
    There are some tv monitor dependent codes in a few routines
    like u(), you should know g(7), g(0) are for the effects
    of reverse video on a vt100-compatible terminal.
*/

char L[99], Inp();
long h[4];
int  new_piece=0;

#ifdef UNIX
t(){   /* set unix timer alarm */
  h[3]-=h[3]/3000;
  setitimer(0,h,0);
}
#endif

int c,d,l,
#ifdef UNIX
v[]={(int)t,0 ,2},
#endif
 w,s,I,K=0,i=FRAME,j,k,q[FRAME],Q[FRAME],*n=q,*m,x=XSTART, xxx,

f[]={ 7,-X-1,-X,1, 8, 1-X,-X,-1,
      9,-1, 1, X,  3,-X-1,-X,-1,
     12,-1,X-1,1, 15,-1,X+1,1,
     /* change to 18,-1,1,2 for a long bar */
     18,-1, 1,0,   0,-X,-1,X-1,
      1,-X,1,X+1, 10,-X,1,X,
     11, -X,-1,1,  2,-X,-1,X,
     13,-X,X,X+1, 14,1-X,-1,1,
      4,-X-1,-X,X,16,1-X,-X,X,
     17,-X-1,1,-1,5,-X,X,X-1,
      6,-X,X,0};  /* change to 6,-X,X,2*X for a long bar */

#ifdef IBMPC

void YourMove(tile)
unsigned int tile;
{ /* move tile to port and pull yourmove signal low */

unsigned int tmpp;

   tile = tile + tile;
          /* Shift tile left one bit. Leave ls-bit = 0. */
   tmpp = (tile+1);
          /* Set least significant bit = 1. **/
```

```
   outportbbb(PortJ, tile); /* Send out in LSByte of port J */
                            /* YourMove is now low. */
   outportbbb(PortJ, tmpp); /* pull YourMove back to high */
   delay(SLEEPTIME);
}

char * port_name(port)
unsigned int port;
{
    static char *name[]={"PortD","PortJ","PortK","UNKNOWN"};
    switch(port){
     case 0x303: return name[0];   /** PortD **/
     case 0x309: return name[1];   /** PortJ **/
     case 0x30A: return name[2];   /** PortK **/
     default: printf("Dont recognize this port %x\n",port);
              return(name[3]);
    }
}

void outportbbb(port, dbyte)
unsigned int port;
unsigned char dbyte;
{
  if(even%2) printf("\n\033[50;42H");
  else printf("\033[50;0H");
  ++even;
   #ifdef SHOWPORTS
     printf("--> outportb: (%2d) %03x to %5s ",
         dbyte/2, dbyte, port_name(port));
   #endif
  outportb(port, dbyte);
}

unsigned int inportbbb(port,silence)
unsigned int port;
int silence;
{
  unsigned int byte;
  #ifdef SHOWPORTS
    if(!silence)printf("\tinportb: reading from %s\n",
                       port_name(port));
  #endif
/* activate this for real game with the computer
   byte = getch();
*/
byte = inportb(port);
#ifdef SHOWPORTS
  if(!silence)printf("\tinportb: just read byte %02x\n", byte);
#endif
```

```
  return byte;
}

void set_up()
{
   /*****************************************************
   *
   *       define fixed addresses of card registers
   *
   *****************************************************/

   PortD = 0x303;               /* Set Port D address */
   PortJ = 0x309;               /* Set Port J address */
   PortK = 0x30A;               /* Set Port K address */

   printf("\n\n\n\n\n\n\n\n\n\n\n\n\n\n\n\n\n\n\n\n");

   outportbbb(PortD, 0x00); /* load reset latch. */
   outportbbb(PortD,0x01);  /* set reset latch to 1. */
}

void get_move()
{
  /* poll the input port until time out
    if time out return -1 to drop the bucket */
   unsigned int mymove;
   unsigned int ro;
   unsigned int co;
   int j,i;
   float timer;

   timer = 0.0;         /* time out if a player thinks too long */
   mymove = 1;

   /* printf("\nI let you think, polling ."); */
   while(mymove && (timer isnt THINKTIME)){ /* printf("."); */

      mymove = inportbbb(PortK,1);
      ++timer;
      mymove = mymove & 0x20;
   }

   if( timer is THINKTIME ){ *(M)=' ';
                             *(M+1)='\0';
                             return;
   } /* drops the tile */

   ro = inportbbb(PortK,1);  /* read rotation */
```

```
                                  /* make sure 0 <= ro <= 3 */
    ro = ro & 0x03;
    co = inportbbb(PortK,1);      /* read lateral translation */
                                  /* make sure 0 <= co <= 7 */
    co = (co/4) & 0x07;

    /*  printf("\n\033[50;40H");
        printf ("Read MyMove = %02x from PortK\n", mymove);
        printf("Read ro = %02x from PortK\n", ro);
        printf("Read co = %02x from PortK\n", co);
    */

    ++co;

    /* now translate the "rotation" and "translation" to
       the appropriate characters */

    j=0;
    while(ro){ *(M+j)='d'; j++; ro--;}

    if(co < XM1 ) {i=XM1-co;
                   while(i){ *(M+j)='h'; j++; i--;}
    } else {       i=co-XM1;
                   while(i>0){ *(M+j)='l'; j++; i--;}
    }
    *(M+j)=' ';       /* drops */
    *(M+j+1)='\0';

   printf("\nScore %4d on level %1d      ",w,l);
   if(ngames>0)
      printf("\nAverage score is %5.1f in %3d games    \n",
             total/ngames,ngames);
   else printf("\nFirst Game !!!!!    \n");
}
#endif

int * types()
{
   int * tile;
   unsigned int drawtile;

   new_piece=1;
   drawtile = rand()%TYPES;
   #ifdef MAN
      drawtile = (getch()-'0')%TYPES;
   #endif
   #ifdef IBMPC
      YourMove(drawtile);
   #endif
```

```
    tile = f+drawtile*4; /* needs to be multiply by 4 due to
                            data structure of f[] */
    return tile;
}

u(){  /* updating the screen */
    for(i=(X-1);++i<BUCKET;)
        if((k=q[i])-Q[i]){
   Q[i]=k;            /* go to (y,x) */
   if(i-++I||i%X<1)printf("\033[%d;%dH",(I=i)/X,i%X*2+28);
            /* print out the coordinates (x,y)
     if(i-++I||i%X<1)printf("(%d,%d)  ",i%X, (I=i)/X);
            */
          printf("\033[%dm  "+(K-k?0:5),k); /* reverse video */
          K=k;
       }
    Q[BUCKET-1]=c=Inp();
}

char Inp()
{
#ifdef IBMPC
    static int move=0;

    if(new_piece){ get_move(); new_piece=0; move=0;}
    if(M[move] != '\0') return  M[move++];
    else return -1;
#endif

#ifdef UNIX
  return(getchar()); /* get character and let it time out
                        if there is no key struck,
                        in this case, return -1

                     */
#endif
}

G(b){
   for(i=SHAPES;i--;)
       if(q[i?b+n[i]:b])return 0;
   return 1;
}

g(b){  /* determine whether it is reverse video 7 and 0 */
   for(i=SHAPES;i--;q[i?x+n[i]:x]=b);
}

main(C,V,a)
char**V,*a;
```

```
{
  h[3]=1000000/(l=C>1?atoi(V[1]):2); /* build the bucket */
   puts("\033[H\033[J");  /* clear screen */

  #ifdef UNIX
     system("stty cbreak -echo stop u");
     sigvec(14,v,0);
     t();
  #endif
  while(ngames<GAMES){

  #ifdef IBMPC
     puts("\033[H\033[J");  /* clear screen */
     set_up();  /* initialize the parallel ports */
  #endif

     srand((unsigned)time((time_t*)NULL));

     for(a=C>2?V[2]:"hdl pq";i;i--)*n++=(i<X2||i%X<2)?7:0;

     for(n=types();  ;g(7),u(),g(0)) {
         if(c<0){
            if(G(x+X))x+=X;
            else{
                g(7);
                ++w;
                for(j=0;j<XY;j=X*(j/X+1))
                    for(;q[++j];)
                        if(j%X==(X-2)){
                            for(;j%X;q[j--]=0);
                                u();
                                for(;--j;q[j+X]=q[j]);
                                u();
                        }
                n=types();
                x=XSTART;
                if(!G(x))c=a[5];
            }
         }
         if(c==*a)G(--x)||++x;                /* right */
         if(c==a[1]){ n=f+4**(m=n);
                  if(!G(x)) n=m;         /* rotate */
         };
         if(c==a[2]) xxx=G(++x)||--x;         /* left */
         if(c==a[3]) for(;G(x+X);++w)x+=X; /* go down */

         if(c==a[4]||c==a[5]) {               /* pause */
            #ifdef UNIX
            s=sigblock(8192);
```

```
                #endif
                printf("\033[H\033[J\033[0m\n");
                if(c==a[5])break;                    /* quit */
                for(j=BUCKET;j--;Q[j]=0);

                while(Inp()-a[4]);
                puts("\033[H\033[J\033[7m");
                #ifdef UNIX
                sigsetmask(s);
                #endif
            }
        }
      printf("\nScore %4d on level %1d\n",w,l);
      Scores[ngames++]=w; total = total+w;
      w=0;
      while(w<ngames){printf(" %d", Scores[w]); w++;}
      printf("\nAverage score is %5.1f in %3d games\n",
            total/ngames, ngames);
      for(w=FRAME-1;w>=0;w--) Q[w]=q[w]=0;
      new_piece=d=c=w=I=K=s=k=j=even=0;
      x=XSTART;
      n=q;
      sleep(1);
      i=FRAME;
    }
    #ifdef UNIX
      system("clear");
      system("tset");
    #endif
    #ifdef IBMPC
      puts("\033[H\033[J");  /* clear screen */
      system("cls");
      printf("Bye Bye. Nice Game! Try again soon.\n");
    #endif
    printf("Score %4d on level %1d\n",w,l);
    printf("\nAverage score is %d in %d games\n",
          total/ngames, ngames);
    return 0;
}
```

Appendix B

A Workview Tutorial

This appendix provides some of the essential steps involved in entering a design using a schematic editor, simulating a design at the logic level, and simulating a design at the timing level.

B.1 Workview

`Workview` is an integrated environment available for the design entry and verification of both Actel and Xilinx FPGAs. The `workview` package consists of facilities for schematic capture, simulation, and netlist translation. The presentation in this appendix is specific to the Xilinx/Workview package.

Logitech Mouse Driver

Some two-button mice and their mouse drivers may be incompatible with the current version of Xilinx software (April 1992). The fix is to use different mouse drivers at different times. Use

```
mouse   PC
```

with `workview`. Use

```
mouse   MP
```

with Xilinx tools such as `xdm`, `xact`.

Where is Workview loaded?

Typically, the package `workview` is loaded in the directory `C:\WORKVIEW`. It's batch file is named "WV." To invoke `workview`, type

```
WV
```

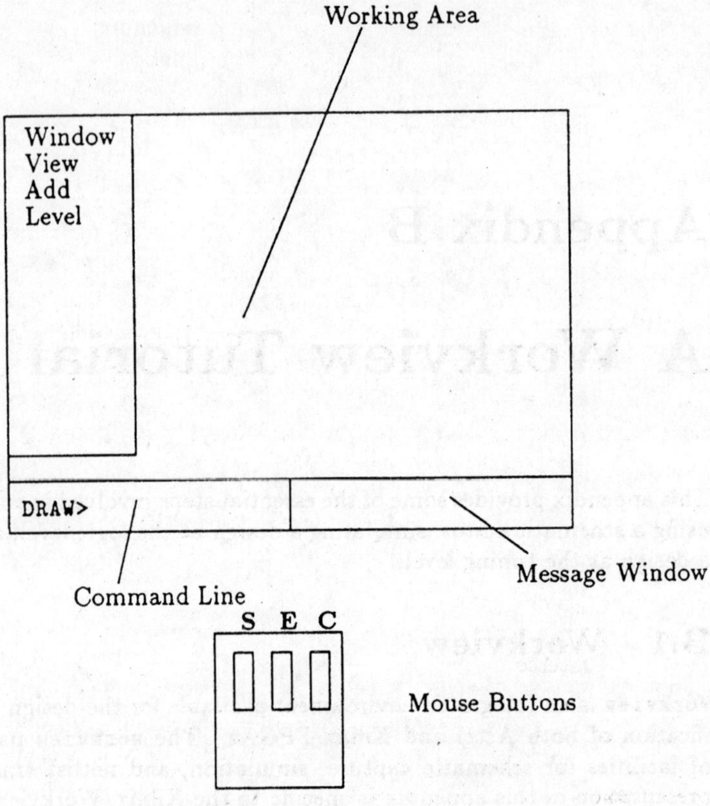

Figure B.1: The Workview Environment.

or

```
workview
```

A file `viewdraw.ini` is read by `workview` when it is evoked. You can customize your own `viewdraw.ini` or the copy in `workview\standard\viewdraw.ini` by keeping your own `viewdraw.ini` file in your working directory. The last few lines in `viewdraw.ini` select the FPGA libraries to be used. For example:

```
DIR 0  .
DIR 12 C:\workview\xttl
DIR 13 C:\workview\x3000
DIR 14 C:\workview\mx3000
DIR 15 C:\workview\builtin
```

If you did it right, you would see a window similar to Fig. B.1. Use the left button of the mouse to Select an object (net, component, symbol, or text), the middle button to Eexecute a command, and the right button to Cancel a command. Finally, just one more thing to remember: use the function key `F6` to pan around your drawing, and use the function key `F7` to zoom in `F8` to zoom out, and `F10` to view the entire design.

`Workview` works mainly with three areas, namely, menu, working area, and the command line.

Help Facility: Help in `viewdraw` may be obtained by typing

```
help <command>
```

in the command line.

Menu: Most of the commands are listed in this screen. The commands may be selected with the help of a mouse. The commands are grouped according to the various functions, Add, File, Quit etc.

Working area: The drawing screen is where you draw your design.

Message window: This area is where `workview` reports messages.

Viewdraw Philosophy

A design in the `viewdraw` schematic editor comprises of "*components*," and "*symbols*" interconnected through "*nets*" or "*buses.*" Plane texts are for documentation purposes. Standard components are defined in the standard macro library. User-defined symbols are in the user's primary directory under `sym`. A symbol may be viewed as a functional block with inherent functional behavior. Pins are the input/outputs for a symbol, very much like a chip, and they act as interfaces between the symbol and the outside world. `Viewdraw` provides standard components A vendor's macro library typically has primitive gates such as NAND gates, NOR gates, inverters, plus FPGA specific components like I/O pads, and special symbols for controlling technology mappings. Apart from the components Library, one may create your own symbols and maintain your own library. As mentioned earlier, like components, symbols are interconnected with nets and buses.

`Viewdraw` basics:

```
Invoking command : workview
Quitting command : quit
```

Graphic elements in `viewdraw`: the various graphic elements are nets, symbols, labels, texts, and windows. You can open multiple windows in `viewdraw`. Nets represent the wires and buses that connect electronic components. Two wires are said to be connected when the connection dot (•) is present.

Symbols represent electronic components. Referring to Fig. B.2, the terminology associated with symbols are:

1. Symbol boundary: outside boundary of the symbol.

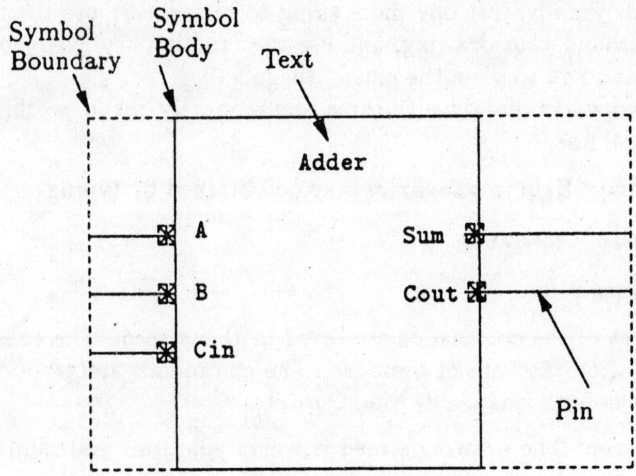

Figure B.2: A Workview Symbol.

2. Symbol body: the main body of the symbol.

3. Pins: the wires connected to the symbol block.

4. Pin labels: names associated with the pins (e.g., **A**, **Sum**).

5. Symbol Id.: text for identification of symbols.

6. Attribute of pin: specify a pin as input or output pin.

A net can be named by giving it a *label*. Plane text has no significant in **viewdraw**, except for documentation purpose. An attribute is a unique property associated with a net, a symbol, or can even be unassociated. Here is a list of important attributes in **viewdraw** using the Xilinx macro library:

```
Attribute Name Value          Description
---------------------------------------------------------------------
  pintype      IN             An input  pin of a symbol
  pintype      OUT            An output pin of a symbol
  EQUATE       F=A            A equation of a Xilinx CLB component
  LOC          P38            The pad number designation of an
                              io pad
  part         XC3020PC84-80  A part number, unassociated
  C                           A critical net
  L                           A net to try use a long line
```

A label has it's *point of effect.* When attaching a label to an object, first select the object, next

Add Label

click the **E** button and then enter the label (string) from the keyboard. In this manner, the label is associated with the object which you selected no matter where the label is. You should label **all** the components, symbols and nets.

B.2 Design Entry in Viewdraw

A typical design entry session in `viewdraw` would involve the following steps:

1. Enter `Viewdraw` drawing screen.

2. Pick the symbols available in the standard libraries, or the ones in your own library. The file `viewdraw.ini` contains lines that define the standard library paths. The following discussion assumes that that we are using the XC3000 library. We shall demonstrate the steps to enter a design as given in Fig. B.3.

3. Open a window to start a new drawing

   ```
   Sch foo
   ```

   Use the **S** button to drag the mouse to define the area of your drawing. Click the **S** button to terminate.

4. To make your schematic to appear really professional, and fool your professor (or boss) into believing that your circuit really works, you should place a title block in your drawing. In `viewdraw`, do

   ```
   Add Comp
   ```

   Click the **E** button to execute the command and then type `asheet` from the key board. Use the middle button to drop the title block at the proper location (preferrably on the lower left hand corner of your window). Put your name and drawing title into the title block by clicking

   ```
   Add Text
   ```

   click the **E** button, and enter the strings from the keyboard.

5. Next, add the required components from the macro library, e.g., load a 2-input AND gate, load the input and output pads, and load the input and output buffers.

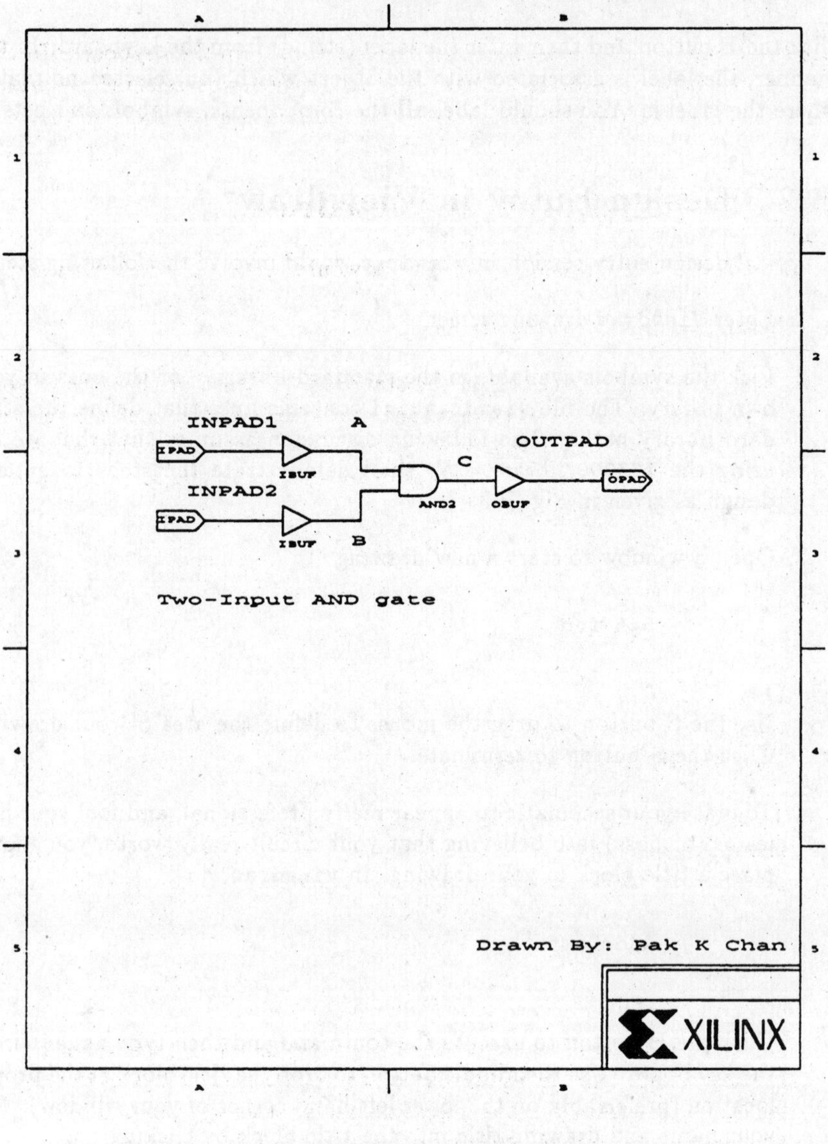

Figure B.3: Design Entry Example.

```
Add Comp and2
Add Comp ipad
Add Comp ipad
Add Comp opad
Add Comp ibuf
Add Comp ibuf
Add Comp obuf
```

Add and place these components one by one by dropping a component using the **E** button. You may duplicate the components from the ones which are on the drawing by using the `copy` facility.

6. If the location of your component isn't right, relocate the component by selecting it with the **S** button and

   ```
   move
   ```

   Use the middle button to drop the component to where you want it to be.

7. Use

   ```
   Add net
   ```

   to connect the components. `Viewdraw` is pretty good in assisting the user to finish the connections as you will notice. Just use the **E** button will help you to finish the job. Click once to select the pin of the component, drag the net with the mouse, click again to make turns, click again to place the net to where you want it to be.

   Other useful commands, just in case you need to make some changes, are:

   ```
   Del Net
   ```

8. Label the nets on the interconnection lines. For example, select the net with the **S** button and

   ```
   Add Label
   ```

   Click the **E** button, then type in the text strings.

9. Save the drawing (`.WIR` extension) by

    ```
    FILE write foo
    ```

10. This will give you a `foo.wir` file in the `WIR` directory, and `foo.1` in the `SCH` directory. Exit.

    ```
    quit
    ```

11. You may edit your change again next time you enter `Viewdraw` by saying

    ```
    sch foo
    ```

12. From DOS prompt, the drawing preprocessor reads `foo.wir` file to create an `XNF` file.

13. Create a Xilinx Netlist Format file for the schematic with the command:

    ```
    wir2xnf -P3020PC84-70 foo
    ```

14. Create an `LCA` format for the schematic with the commands:

    ```
    xnfmap  foo
    map2lca foo
    ```

15. Place and route the `LCA` file with `apr`.

    ```
    APR foo foor
    ```

16. View the placed and routed design `foor` with `xact`.

You may automate this sequence of commands in a "make" file providing that you have the make facility available.

Printing Schematics

To use a postscript printer to print out a schematic, you need to generate a postscript file of the schematic. In `workview`, select

```
Plot Device PS
```

Then

```
Options File
```

Type `foox` as file name and click

```
Go
```

Use defaults such as sheet, `A` size paper, and 0 rotation.

Sources of information

1. `Viewdraw` user manual. ViewLogic, Corporation.
2. XACT Development System Vol II. Chapters 2 and 5.
3. XACT LCA MACRO Library.
4. Workview Tutorial, Xilinx User Guide and Tutorials 1991.

Functional Blocks

A functional block is a box (module or symbol) that represents a portion of the circuitry in a design.

In the following exercise, we shall create a 2-input AND functional block as a **module**. The content of the functional block originates from an equation file `and.eqn`:

```
INORDER = a b;
OUTORDER = c;
c = a*b;
```

We convert the `eqn` file format to `XNF` file format using

```
%>  eqn2xnf odd.eqn
```

The XNF `odd.xnf` file contains the lines:

```
LCANET,2
PROG, EQN2XNF, Ver 1, Computer Engineering @ UC Santa Cruz
PART, 3020PC84-50

SYM, c, AND
PIN, O, O, c
PIN, 1, I, a
PIN, 2, I, b
END
EOF
```

The next step is to draw a *top-level* schematic diagram `top` in `workview`, as illustrated in Chapter 3, Fig. 3.11. The functional block `ODD_FUNC` is defined as a module (not a composite). (When you save your symbol drawing: use `Change Block Type Module`.) The attribute `FILE=odd` must also be added to the module.

To merge the `odd.xnf` file with the rest of your drawings, do

```
wir2xnf -P3020pc84 top
xnfmerge top flat
```

before you do the technology mapping, placement and route, which is

```
xnfmap flat
map2lca flat
apr flat flat_route
```

For entering state machines, draw the D flip-flops (FD) from the library in **workview**, and include the combinational part of the state machine as a functional block using the technique previously mentioned.

Hints and Guidelines

- Use Level, Push or Pull to view schematic drawings at different hierarchical levels.
- Label *all* nets.
- Use the global clock buffer, GCLK (XC2000 and XC3000 libraries), in the library to drive all clock signals.
- Specify all the input pads, output pads, buffers, and global clock buffers at the top-level schematic diagram.
- **Workview** doesn't like symbol names with square brackets, unlike **misII**! Use a familiar text editor to carry out some character translations.

Functional/Logic Simulation

After entering and saving a schematic drawing **top** in **workview**, say

```
Export Wirelist Viewsim
```

This will create a **top.vsm** netlist ready for logic simulation.

Select a group of nets that you would like to observe and control by using the left button **S** to click the first net and use the right button **C** to select the rest of them.

Bring up the simulator **viewsim** by

```
Open Window Viewsim
```

Open the window and then select it, you will see **viewsim active** at the lower left hand corner of the window. Establish the connection between the program **viewwave** which is responsible for displaying the waveforms by

```
Setup Vwave
```

You will be prompted for files and nodes, etc. Just use the defaults (**top.vsm** by clicking the middle button.

You are now ready to simulate, go to the **viewsim** window, commands to control the simulations are:

```
h <netname>
l <netname>
step <stepsize>
clock <clk> 0 0 1 1
cycle <number of clock pulses>
sim <steps>
restart
```

A sample sequence to initialize the inputs and then simulate for 5ns is

```
h input1
l input2
sim 50
```

You can update the waveforms now by selecting the **viewwave** window. You can also back-annotate the logic values to the schematic by selecting the **viewdraw** window.

Global Reset

A special signal **gr** (not **globalreset-**) is available for global reset to initialize all the Xilinx flip-flops to zeros. It is an active low signal. This is the only way that you can initialize all the flip-flops.

Simulation Command Log and Script

A log file of all the commands that you used is kept in **viewsim.log**. You can use this file (or a variant of it) to rerun/repeat the simulation, by typing **execute myscript** in **viewsim**.

Timing Simulation

This is essentially the same as functional simulation, except that there are a few more steps involved in back annotating the timing information to your schematic drawing. Suppose that you have a place-and-routed design **top.lca**. Following the given steps:

```
lca2xnf top top_r
bax top_r top.map -o top_bann
xnf2wir top_bann
vsm top_bann
```

This will generate a back-annotated (with timing information) **vsm** file **top_bann.vsm**. Use this master simulation file when you open the **viewsim** window.

Multiple FPGAs

The vanilla version of **viewsim** doesn't support simulation of multiple FPGAs design in the timing mode, however, you can "cheat" in the logic mode. Just put all your components together into a big schematic drawing.

Functional blocks

Since **viewsim** doesn't read **XNF** format, functional blocks entered as **XNF** files in **viewdraw** cannot be simulated directly. To circumvent the problem, generate a flattened **xnf** netlist using **wir2xnf** and **xnfmerge**, e.g., then use **xnf2wir** and **viewgen** to produce a flattened schematic.

Index